THE HEAT IS ON

THE HEAT IS ON

The High Stakes Battle over Earth's Threatened Climate

ROSS GELBSPAN

ADDISON-WESLEY PUBLISHING COMPANY, INC.
Reading, Massachusetts Menlo Park, California New York
Don Mills, Ontario Harlow, England Amsterdam Bonn
Sydney Singapore Tokyo Madrid San Juan
Paris Seoul Milan Mexico City Taipei

Library of Congress Cataloging-in-Publication Data

Gelbspan, Ross.
The heat is on : the high stakes battle over Earth's threatened climate / Ross Gelbspan.
p. cm.
Includes bibliographical references and index.
ISBN 0-201-13295-8
1. Global warming—Government policy. 2. Energy Industries—Political activity. 3. Greenhouse effect, Atmospheric—Government policy. 4. Global warming—Government policy—United States. 5. Energy industries—Political activity—United States. 6. Greenhouse effect, Atmospheric—Government policy—United States. I. Title.
QC981.8.G56G45 1997
363.738'74—dc21 96–37242
CIP

Jacket design by David High
Text design by Karen Savary
Set in 11.5-point Adobe Garamond by Pagesetters, Inc.

123456789-MA-0100999897
First printing, April 1997

This book is dedicated to Thea and Joby, to Jake and Mike and Joshua and Sahar and Hajar and Charlotte and Sam and Nikki and Ibrahim, Khalim and Mohammed and Nicolas, Ben and Molly and Peter and Eric and all the other children who have decided—or will be deciding—to make this earth their home.

CONTENTS

INTRODUCTION

Climate Change Is Here. Now.

IN JANUARY 1995 A VAST SECTION OF ICE THE SIZE OF Rhode Island broke off the Larsen ice shelf in Antarctica. Although it received scant coverage in the press, it was one of the most spectacular and nightmarish manifestations yet of the ominous changes occurring on the planet.

As early as the 1970s, scientists predicted that the melting of Antarctica's ice shelf would signal the accelerating heating of the planet as human activity pushed the temperature of the earth upward. They were not wrong.

Two months later, a three-hundred-foot-deep ice shelf farther north collapsed, leaving only a plume of fragments in the Weddell Sea as evidence of its twenty-thousand-year existence.

Dr. Rodolfo del Valle, an Argentine scientist who witnessed the disintegration of the ice shelves, had little doubt about the implications when he recalled the events to a reporter from Reuters. "If conditions remain unchanged," he said, rising global temperatures "could cause catastrophic flooding all over the world. We thought the flooding would occur over the course of several centuries, but the whole process has been much quicker than we anticipated."

The first indications of the massive fissures had occurred in mid-January, when del Valle was working at a research station on James Ross Island, just off the Antarctic peninsula. Colleagues at an Argentine base on the Larsen ice shelf radioed him there to say that they were being shaken by constant ice quakes. A few days later they called with even more startling news.

"On the twenty-third of January, they called me over the radio and said: 'Rudi, something's happening, the ice shelf is breaking.' An enormous crack had opened from the edge of the shelf on the Weddell Sea up to the mountains," said del Valle, who heads the earth sciences department at Argentina's National Antarctic Institute. Flying six thousand feet overhead in a light plane, del Valle saw that the ice shelf, up to a thousand feet thick in places, was beginning to break up into smaller icebergs. "It was spectacular because what once was a platform of ice more than forty miles wide had been broken up into pieces that looked like bits of polystyrene foam . . . smashed by a child.

"The first thing I did was cry."

The second monstrous crack—a forty-mile-long fissure that occurred two months later—appeared in the northernmost part of the Larsen shelf, which runs six hundred miles up the Antarctic peninsula. The cracking was caused by a regional warming that began in the 1940s. Measurements in the Antarctic peninsula show that its average temperature has risen by nearly 20 degrees Fahrenheit in the last twenty years. Del Valle said that several months earlier he and his colleagues predicted that the ice barrier would crack in ten years. "But it happened in barely two months," he observed. "Recently I've seen rocks poke through the surface of the ice that had been buried under six hundred meters of ice for twenty thousand years."

Ice works like a mirror. Its surface reflects most of the heat radiated by the sun back into the atmosphere. As ice like the Larsen ice shelf breaks up and newly exposed rocks absorb heat and further warm the ice cap, the earth will reflect less heat and retain more of it.

While the breakup of the Antarctic ice shelf may be the most dramatic manifestation of global warming, other symptoms have been emerging around the world for several years.

While the experts supposedly debate whether global warming has been "proved," the recent weather record has been marked by more intense rainfalls and severe snowstorms; stronger hurricanes; unseasonal and prolonged droughts, resulting in crop failures and devastating forest fires from Texas to Mongolia; killer heat waves from India to Chicago; and record flooding from Yemen to Nepal.

All of these escalating climate extremes share a common source, according to the overwhelming consensus of scientific thought: they are nature's expressions of the early stages of the heating of the atmosphere.

But some individuals do not want the public to know about the immediacy and extent of the climate threat. They have been waging a persistent campaign of denial and suppression that has been lamentably effective.

Four months after the fracturing of the Larsen ice shelf, the Washington-based, ideologically conservative George C. Marshall Institute released its fourth consecutive report emphatically dismissing the notion of planetary warming. "Comparable temperature changes are commonplace in recent climate history," according to the institute, which conducts no original research itself and whose reports are viewed by the vast majority of scientists as political statements rather than as research contributions. Its own chairman has acknowledged that the institute's reports represent nothing more than opinion.

But in some circles the opinions of the Marshall Institute carry more weight than those of the world's leading scientists. In May 1996 the House Science Committee was about to make its final budget decisions for the year. Before that meeting, the committee chairman, Robert Walker, a Republican congressman from Pennsylvania, asked the National Research Council (NRC) whether Congress should continue to fully fund NASA's Mission to Planet Earth, a program designed to monitor changes in the global climate and their effects. The NRC strongly recommended full funding for the program. Walker then turned to the Marshall Institute for advice. When he successfully recommended cutting the funding for the program, he cited the Marshall Institute's denials of the climate crisis to justify his decision.

It was not Representative Walker, however, who controlled the front line of the climate battle in the 104th Congress. In 1995 Dana Rohrabacher, a dapper, small-town journalist turned Republican congressman from Orange County, California, presided over a series of hearings on climate change in a House subcommittee. Discussing sea level rise, a consequence of atmospheric heating that scientists estimate will reach three feet within the next century, an Environmental Protection Agency (EPA) official told Rohrabacher that such a rise "could drown [up to] 60 percent of our coastal wetlands." That rise in sea level, he said, "would inundate more than five thousand square miles of dry land in the United States, an area the size of Connecticut," and some of its consequences would be "potentially catastrophic and irreversible."

The 49-year-old, round-faced Rohrabacher, an avid surfer, was unmoved by the gravity of the situation. "I am tempted to ask," he responded, "what this will do to the shape of the waves and rideability of the surf. But I will not do that. I will wait until later, when we get off the record."

After hearing from some of the world's leading climate scientists—who testified that global warming has already begun and that its early symptoms are already upon us—Rohrabacher, who chaired the Subcommittee on Energy and Environment of the House Committee on Science, accused them of practicing religion rather than science. He proceeded to vote to defund several government research programs that monitor the escalating stresses on the global environment.

In so doing, the congressman accomplished a staggering assault on foundations of our scientific knowledge. But he could never have succeeded in pulling off this legislative equivalent of a book burning were it not for the unique power of the one interest his leadership was benefiting: the fossil fuel industry. In dismissing the findings of the world's leading scientists about climate change, Rohrabacher was carrying out the strategy of big oil and big coal—

and oil and coal emissions are the major driving force behind the heating of the planet's atmosphere.

In the United States the truth underlying the increasingly apparent changes in global climate has largely been kept out of public view. As a result, what most Americans know about global warming is obsolete and untrue.

That is no accident. The reason most Americans don't know what is happening to the climate is that the oil and coal industries have spent millions of dollars to persuade them that global warming isn't happening.

Together, oil and coal constitute the biggest single industry in history. Big oil alone does well over a trillion dollars a year in business. Coal and oil products are used by every home, every car, every factory, and every building in most countries of the world. Moreover, the invisible and well-funded public relations apparatus of the oil and coal lobby reaches many, many minds.

Denying global warming is a clever strategy—but its shortsightedness carries the seeds of massive destruction.

In November 1995, 2,500 leading climate scientists announced that the planet is warming because all the emissions from coal and oil burning are trapping in more of the sun's heat than is normal for our climate. Even if that warming is not yet obvious, they warned, it is already generating bizarre and extreme changes in the weather. This new period of less stable climate we have entered, the scientists wrote, "is likely to cause widespread economic, social and environmental dislocation." Their report noted that "potentially serious changes have been identified, including an increase in some regions of the incidence of extreme high temperature events, floods, and droughts, with resultant consequences for fires, pest outbreaks and ecosystem[s]."

This is a very frightening prospect. Starting slowly, we may already have embarked on a profoundly horrifying journey. Almost every week international news wires carry stories about extreme, disruptive, and often record-setting weather events. Unfortunately, these stories normally get buried away in local newspapers, and powerful forces are at work to see that they remain obscure and their significance unrecognized.

In the late summer of 1995 the immediacy of climate change struck me as I was sitting on my ramshackle back porch outside Boston. What was then a wilted patch of straw had once been my lush, green lawn before that summer, which was one of the driest in New England history. Overhead, between my porch and the far boundary of the atmosphere twelve miles up, carbon dioxide, driven upward by the unceasing flow of emissions from burning coal and oil, continued its relentless increase. This invisible gas is gradually warming the planet by trapping the sun's heat and preventing it from being reflected back into space. Higher up, just beyond the boundary of our atmosphere, water vapor continued silently to accumulate, amplifying the heating of earth. Hot as it was as I sat on my porch, the temperatures on earth could have been even hotter: Some of the warming of the atmosphere was masked by an invisible umbrella, about five miles overhead, of sulfate pollution, which deflected the sun's heat back into space. But that cooling effect, I knew, was only temporary.

My porch faces east. Two miles dead ahead the Atlantic Ocean was rising at a faster rate than at any time in human history, approaching a tenth of an inch a year. While scientists caution that this rapid rate may be temporary, the general trend toward significantly higher sea levels is not. One-tenth of an inch seems like a small amount until you stop to think that the oceans cover more than 70 percent of the earth's surface. A sea level rise of even a few inches would wreak havoc on many of the world's coastal communities.

Beyond the Boston Harbor islands, I mused, beyond the horizon, on the far side of the Atlantic, alpine plants are migrating up the mountainsides of Europe, seeking ever higher altitudes to survive the rising temperatures. Two thousand miles south, people living in the mountains of Rwanda and Kenya suffer from malaria and yellow fever, brought by mosquitoes that have recently migrated from lower levels. Only a few years ago, these mountain areas were free of such disease-carrying insects, because it was too cold for their survival there.

Environmentalists have been warning us about global warming for years. What is news is that global warming is no longer merely a future possibility. The heat is on. Now. Its early impacts, in the form of more extreme and unstable climate, are being felt even as I write.

The purpose of this book is not only to bring home the imminence of climate change but also to examine the campaign of deception by big coal and big oil that is keeping the issue off the public agenda. I will examine the various arenas in which the battle for control of the issue is being fought—a battle with surprising political alliances and infuriating episodes of relentless obstructionism. The story told here provides an ominous foretaste of the most likely political consequences of our continued denial of the gathering threat—political chaos and the spread of totalitarianism. And it will conclude by outlining a transition to the future that contains at least the possibility of continuity for our existence as an organized, civilized society.

Sitting on my porch, remnants of my daughters' childhood were still visible in the parched yard. The weather-shredded marker where they buried their beloved cat. The thick screw hooks in the four-trunked poplar tree where they once hung a hammock. The same sun that dried my yard has also parched the eroded cornfields in the Andes Mountains of southern Ecuador, where my oldest daughter worked in a remote village the previous summer. About five hundred miles south of that village of Cashapugro, the ancient mountain glaciers of Peru are retreating at a frightening rate.

The late summer afternoon warmth stirred lazy reveries of my own approaching old age. Along with the genetic material and psychological baggage of my wife and myself, our daughters, now in college, carry our deepest hopes for the future. I had once fantasized a world of peace, in which my grandchildren would thrive in secure and nurturing communities. But that daydream has been overwhelmed by my terrible knowledge of a reality that is advancing with a gathering vengeance.

The problem is not difficult to understand. Each year humans pump 6 billion tons of heat-trapping carbon into the lower atmosphere, which is only twelve miles high. Within a few decades the atmospheric levels of carbon dioxide will double from preindustrial levels. The world's temperature record is already bearing witness to global climate change: Since 1980 we have seen the ten hottest years in recorded history. The five hottest consecutive years on record began in 1991. The hottest year in the world's recorded weather history was 1995. The planet is warming at a faster rate than at any time in the last ten thousand years.

It is a problem of mind-boggling magnitude, marching inexorably toward a climax. "If the last 150 years had been marked by the kind of climatic instability we are now beginning to see," says one world-class scientist, "the world might never have been able to support its current population of five billion people."

As horrifying as it is, the nature of the problem is not mysterious; nor is its cause. But another part of the problem is harder to fathom. It is taking place offstage behind closed doors, where the affairs of state and industry are conducted. In the media, where public opinion is formed; in the halls of Congress, where laws are made; and in international climate negotiations, the facts of global warming are being denied with a disinformation campaign as ferocious as any in history.

A major battle is under way: In order to survive economically, the biggest enterprise in human history—the worldwide oil and coal industry—is at war with the ability of the planet to sustain civilization. The trillion-dollar-a-year coal and oil industry is pitted against the oceans, forests, ice caps, and mountains of the earth as we know them today.

With sales exceeding two billion dollars every day, the oil industry supports the entire economies of most of the nations of the Middle East and large segments of the economies of Russia, Mexico, Venezuela, Nigeria, Indonesia, Norway, and Great Britain, among others. It employs millions of workers and affects the fortunes of tens of millions of investors. Substantial cuts in oil and coal consumption could, according to some economists and oil industry studies, disrupt entire nations and cripple the global economy.

Eventually, since the unstable climate poses threats even more dire than these to the survival of civilization, we will have no choice but to replace coal, oil, and eventually natural gas with other energy sources. This will mean the extinction of the fossil fuel industry. Nevertheless, the coal and oil lobbyists figure that they can buy time against that inevitable extinction by mystifying the public, casting doubt on every new piece of evidence that confirms global warming. They have good financial back-up, not only in their own immense war chests but in the vast wealth of the oil-producing nations of the Persian Gulf.

To carry out their mystification campaign, the industry's public relations specialists have made extensive use of a tiny band of scientists whose views contradict the consensus of the world's experts. The deep-pocketed industry lobby has promoted their opinions through every channel of communication it can reach. It has demanded access to the press for these scientists' views, as a right of journalistic fairness. Unfortunately, most editors are too uninformed about climate science to resist. They would not accord to tobacco company scientists who dismiss the dangers of smoking the same weight that they accord to world-class lung specialists. But in the area of climate research, virtually no news story appears that does not feature prominently one of these few industry-sponsored scientific "greenhouse skeptics."

It is a most impressive campaign, this assault on our sense of reality. It has carried the day in Congress, where a group of highly ideological legislators successfully waged a war on science. It has been a potent weapon on the international stage, permitting the corporate coal and oil interests—in tandem with the Organization of Petroleum Exporting Countries (OPEC) and other coal and oil exporting nations—to frustrate diplomatic attempts to address the crisis meaningfully.

Most importantly, the campaign has exerted a narcotic effect on the American public. It has lulled people into a deep apathy about the crisis by persuading them that the issue of climate change is terminally stuck in scientific uncertainty.

It is not.

For the last few years, the ravages of the changing climate have been sending shock waves through the executive suites of insurance companies in Zurich, Washington, Munich, and London. The recent succession of floods, hurricanes, and other extreme weather disasters, if they continue as a long-term trend, could bankrupt the insurance industry as such, and these executives know it.

In the relatively calm decade between 1980 and 1989, the insurance industry paid out, on the average, less than $2 billion a year for nonearthquake, weather-related property damage. But in just the five years between 1990 and 1995, the hurricanes, cyclones, and floods in Europe, Asia, and North America have cost the industry an average of more than $30 billion a year, according to the large German-based reinsurance firm Munich Re.

Today insurers worry that a direct hit on, say, Miami or New Orleans by a warming-intensified hurricane could leave them with $50 billion in insured losses. In 1995 Andrew Dlugolecki, a manager for the United Kingdom branch of the insurer General Accident, told a conference of insurers, bankers, and pension managers that "there is no doubt that weather patterns are changing. There is also no doubt . . . that small climate changes have very big effects on society. . . . Looking forward, I am sure that climate change will speed up. I am also sure that there will be major consequences for insurers."

Others in the corporate world, outside the fossil fuel industry, are also beginning to realize that new climate extremes could abruptly curtail the growth of the global economy. Their concerns do not derive from altruism. They are not in the business of environmental protection. What motivates them is a desire to protect their own industries or to increase their current market share. But as they mobilize in opposition to the relentless efforts of big oil and big coal to protect their commerce at any cost, the emerging alliances between insurers and non-oil big business hold significant promise as a positive force for large-scale change.

The spokesmen for big oil and big coal claim that "greenhouse hysteria" threatens to undermine the world's economic foundations. Taking action to reduce atmospheric warming, they say, would plunge humanity into a regressive and dramatically impoverished state. Even conceding the scientific validity of the climate threat, they say, would be the first step down a slippery slope toward desolate poverty and the living standards of our cave-dwelling ancestors.

But the consequences of planetary warming are even more terrifying—a cascade of environmental effects that could shake our civilization off its foundations altogether.

The catalog of anticipated effects is familiar—it reads like the biblical apocalypse. Scientists say these consequences will include not only more extreme temperatures, with hotter heat and colder cold, but also more intense rain and snowstorms, extraordinarily destructive hurricanes, and protracted, crop-destroying droughts, particularly in the interior regions of continents. Island nations and low-lying coastal regions everywhere might disappear under rising seas. These ecological shifts would trigger outbreaks of infectious diseases, as they have already begun to do.

If this scenario isn't hair-raising enough, researchers warn that the feedback effects of global warming could fuel even more nightmarish ones, in which initial warming would cause changes that would themselves intensify and, in turn, speed the warming even more. For example, higher temperatures could lead to drought and wildfires that would burn vast areas of forest. Since vegetation acts to absorb carbon dioxide, that burning would accelerate the accumulation of greenhouse gases—leading, in turn, to more warming-driven climate destabilization. As the greenhouse gases accumulate, the stress could catapult the global climate system into an epoch of wild instability, accompanied by a further rapid and uncontrollable surge of heating.

These feedback effects could cause more than ecological dislocations. They could ripple, as well, through our social fabric.

However insulated we may seem from nature, our planetary environment shapes the way we live. Our interactions reflect the variety of its geography; our institutions mirror its abundances and its stresses. Under the pressure of intensifying instabilities, climate changes and their feedback effects could very likely overwhelm the foundations of democracy. They could well mean a return to the age-old plague of totalitarianism.

In poorer and more environmentally vulnerable countries, the antidemocratic threat is easy to foresee. One of the first casualties of a succession of ecological disasters would likely be individual liberties and governmental democracy and their replacement by a permanent state of eco-emergency.

In wealthier countries like the United States, the impacts of a changing climate are very likely to catch both government and the corporate sector unprepared. And since carbon dioxide lingers in the atmosphere for one to two hundred years, it will continue to disturb the global climate long after we drastically cut our fuel emissions—whether we do so voluntarily or involuntarily. By the time we actually feel the heavy brunt of climate-driven catastrophes, it may well be too late for us to preserve any semblance of democratic order. Governments would likely be forced to resort to martial law in order to respond to weather-related homelessness, disease outbreaks, food shortages, and economic disruptions.

Add the advent of totalitarianism to the scenario of a disturbed climate and the collapse of much of the world's economy, and you get a sense of the stakes of the battle.

There is, moreover, a further complication to this conflict—the yawning chasm of economic inequality between rich and poor. It is an issue of morality to some and irrelevant to others. But in the context of earth's changing climate, the consequences of global economic inequality are as critical as the carbon in the air. Any solution to climate change must address this economic gulf. Regardless of whether it holds a place in one's moral or economic accounting system, addressing it is a critical prerequisite of any solution. Any cutback in oil and coal use by the wealthy countries will be dwarfed by a surge in carbon dioxide production—not to speak of the continuing destruction of the world's CO_2-absorbing forests—by China, India, Brazil, Mexico, and others. Unless they are provided with renewable, climate-friendly energy sources, the developing economic giants will likely be forced to accelerate their burning of fossil fuels to survive the relentless expansion of that poverty, driven by their expanding populations.

On the front of international diplomacy, the large developing nations have viewed with suspicion any initiative that might limit the supply of energy they need to feed and house and employ their citizens. At the same time, the oil- and coal-producing nations have been very successful in obstructing attempts to reduce the world's use of coal and oil. But even in the face of this resistance, an unlikely coalition is prodding the world's governments into the beginnings of meaningful action. A number of small island nations—stretching from the Philippines to Jamaica—fear being flooded out of existence by rising sea levels and increasingly severe hurricanes; they have found common cause with Germany, Britain, Ghana, and other nations that fear the disastrous consequences of unchecked climate change. And they are beginning to force the world's negotiators to confront the problem.

Then there is this perversely good news: As the crisis becomes increasingly severe, the intensity of its destructive impacts will force us to overcome the current plague of disinformation—and the resistance of the deepest-pocketed, biggest private interest in history—to address the problem.

Despite the gravity of this cosmic squeeze we are in, solutions are within our reach. Creating a new energy economy is not a question of science; neither is it a question of technology. Armies of inventors and engineers, fascinated by the challenges of energy, have already supplied us with the devices and designs we need.

For hundreds of years, the engines of economic production have been powered by carbon-based fossil fuels. But in the last few decades, coal and oil have themselves turned into humanity's greatest unintended weapon of mass destruction. It is a happy coincidence of history that at the same time, the climate-friendly energy technologies that we need in order to begin to stabilize the atmosphere have already been created. We have at hand all the technology and knowledge we need to replace the world's supply of coal and oil and, eventually, of natural gas. They need only to be taken to the level of mass production and deployed throughout the world.

The escalating alarm signals emanating from our increasingly disturbed planet contain the seeds for a sweeping transformation. It is not hard to understand that nature is holding a gun to our head. What is harder to understand, at first, is that this moment also presents an opportunity for extraordinary positive change.

Historically, our energy transitions have occurred by accident. This time, as global citizens responsible for a common future, we can actively plan and manage the transition into our next energy epoch. What it requires is taking control of a natural resource-turned-hazard—before it trashes our planet and truncates our history.

ONE

Of Termites and Computer Models

In May 1995, following New Orleans's fifth consecutive winter without a killing frost, the city was overrun by mosquitoes, cockroaches, and termites. "Termites are everywhere. The city is totally, completely inundated with them," said Ed Bordees, a New Orleans health official, who added that "the number of mosquitoes laying eggs has increased tenfold." Bordees attributed the infestation to the lack of frost, combined with unusually high levels of rainfall—it totaled 80 inches in the previous year.

Across the Atlantic, residents in the area around Cádiz, in southern Spain, were suffering through the fourth year of the worst drought in that country's recorded weather history. Until 1992 the region had had the highest amounts of rainfall in Spain—84 inches a year. Since then, it has dropped by more than half, to 37 inches.

On June 2, 1995, Russian thermometers soared to 93 degrees, melting chunks of asphalt at Sheremetyevo airport. It was the hottest June 2 in Moscow since 1889. Just over a month earlier, snowflakes had covered the Russian capital during the coldest April in the previous 120 years of record-keeping.

In China at the end of May, unseasonal torrential rainstorms in Sichuan province killed 62 people, left 25,000 homeless, and destroyed $30 million in property. Two weeks later, Lin Erda, director of China's Academy of Agricultural Sciences, warned his country to prepare for increasingly severe and prolonged droughts, intensified typhoons, and erratic rainfall. If China's climate trends continue, he added, much of the country will face shorter crop-growing periods and increased water deficits in the next century.

In mid-June 1995 the death toll from a broiling heat wave in northern and central India reached 300. Temperatures topped 113 degrees Fahrenheit in Uttar Pradesh, Rajasthan, and other regions. Press accounts noted that Indian summer heat waves are normally broken by intermittent rainstorms. But that year there was no respite from the relentless heat.

Halfway around the globe, in the midwestern United States, a series of rainstorms triggered the second 100-year flood in the region in three years. A month later, a heat wave in Chicago killed about 500 people.

Although it has lurked in the dim margins of public attention for the last few years, global warming first emerged on the public stage during the brutally hot summer of 1988, when Dr. James Hansen of NASA's Goddard Institute for Space Studies warned a congressional panel that it was at hand.

Following Hansen's testimony, growing numbers of scientists raised more concerns about warming, predicting erratic and extreme weather events—early symptoms of a rise in the average global temperature. (Today the term *climate change* has emerged as the name of choice for the warming-driven destabilization of the planetary climate system.) An extreme possibility that many scientists fear is that, under increasing stress, the earth's climate could snap abruptly into a much warmer regime. But even in the absence of such an abrupt, catastrophic change, they fear that the heating could produce increases in severe floods that threaten coastlines, and a wave of droughts, hurricanes, and snowstorms. They foresee rising sea levels and the spread of infectious diseases unleashed by warmer temperatures. What ultimately troubles a consensus of scientists is the prospect of a dramatic rise in the average global temperature on the order of 3 to 4 degrees Celsius (6 to 8 degrees Fahrenheit) in the next century, if present trends continue. That change is roughly equal in magnitude to the difference between the last ice age and the climate today.

In 1988, in response to those concerns, the United Nations established the Intergovernmental Panel on Climate Change (IPCC) to assess the impacts of climate change. The assessments were to be used as the basis for strategies by which nations could curtail their CO_2 emissions. Comprising the leading 2,500 relevant scientists in the world, the IPCC issued a series of reports through its two primary working groups. Working Group I focused on the state of the science while the mission of Working Group II was designed to analyze the impacts of climate change. (A third IPCC panel, composed primarily of economists, was commissioned to examine primarily the costs of mitigating climate change.) The two IPCC scientific panels set up to report to the UN Framework Convention on Climate Change constitute one of the most authoritative bodies of scientists ever assembled around a single area of scientific concentration. In 1990, in the first of four reports the panel has issued, the IPCC concluded that a doubling of carbon dioxide near the middle of the next century will increase the average global temperature by as much as 4.5 degrees Celsius. (They subsequently revised that upper limit down to 3.5 degrees, not because of any scientific correction, but, ironically, because of estimated increases in emissions of low-level air pollutants. In the short term, these pollutants mask the warming of the larger atmosphere.) In 1994, the IPCC, citing new findings from ancient ice-core records, noted that the planet's temperature is demonstrably sensitive to changes in CO_2 concentrations in the atmosphere. The paleontological record, as interpreted by the scientists, shows that prehistoric changes in carbon dioxide concentrations correlate very closely with rapid, dramatic snaps in the climate.

You don't have to be a scientist to understand the basic workings of the greenhouse effect. Under normal conditions, when the sun's rays warm the earth, a percentage of that heat is reflected back into space. The rest of the heat is absorbed by the oceans and the soils and warms the surrounding areas to create the climate conditions we live in. But the recent buildup of carbon dioxide in the atmosphere traps in heat that otherwise would be reflected back into space. The resulting warmth expands ocean water, causing the sea level to rise—just as the level of water heated on a stove rises in its pan. The heating also accelerates the process of evaporation, even as it expands the air to be able to hold more water. The resulting airborne water vapor, in turn, traps more heat, perpetuating the cycle. The more heat that is trapped, the more intense the greenhouse effect.

On the planetary scale, the earth's carbon cycle is at the center of the evolution and continuation of life. Soon after the earth's formation, its atmosphere consisted of around 95 percent carbon dioxide. As plant life evolved, the spreading vegetation absorbed that carbon dioxide and, as it decomposed, stored it in the form of carbon-rich coal and oil deposits. Eventually the earth's plants and oceans absorbed and stored so much carbon dioxide that they reduced the atmospheric concentration of the gas to less than three-tenths of one percent—a level that made the earth's climate hospitable to oxygen-breathers, mammalian life, and human civilization. But the recent explosion in fossil fuel emissions has begun to increase that CO_2 concentration—in effect, reversing millions of years of the natural carbon cycle.

In terms of the carbon content of our atmosphere, "we are in uncharted waters relative to the last 400,000 years," according to Dr. James McCarthy, who is director of Harvard University's Museum of Comparative Zoology and a professor in Harvard's department of earth and planetary sciences.

In June 1995 more than 70 people died from flooding in Bangladesh, while countless others suffered from outbreaks of malaria as mosquitoes swarmed in following the heavy rains. Overall, the floods affected nearly 10 million people. "The area

> *suffers flash floods almost every year, but they [normally] never last more than two or three days," said a provincial official, who noted that this year's rainstorms lasted nearly two weeks.*
>
> *The following month, Ghana experienced its heaviest rainfall in thirty years.*
>
> *That same summer, according to the British Meteorological Office, Great Britain endured its hottest summer since 1659 and its driest summer since 1721.*
>
> *In July northeastern Brazil suffered its worst drought in this century, as rainfall in the region declined by 60 percent. Nine months later, mudslides there, triggered by torrential rainstorms, killed 26 residents. "In the last twelve hours, we have received 9.3 inches of rain when precipitation for the whole month averages fourteen inches," said the mayor of the provincial capital of Salvador.*

Despite the international scientific consensus, the small band of industry-sponsored "greenhouse skeptics" continue to maintain that the trend is too unproven to justify action. Until recently, their skepticism had some basis, given the black holes of uncertainty in our knowledge of the planet's complex and exquisitely interrelated climate system. The skeptics have repeatedly pointed out, for example, that although the world's output of carbon dioxide has essentially exploded since 1940, there has been no corresponding increase in the global temperature since that time. They argue, moreover, that scientists know too little about certain dynamics—ocean-atmosphere exchanges, the role of clouds, the mechanisms of the deep oceans' transfer of heat from equatorial to other latitudes—to predict accurately the behavior of the global climate. The naysayers also emphasize the inadequacy of a major climate research tool—a type of computer model known as the general circulation model. This model is technically unable, they argue, to perform the large numbers of calculations needed to completely simulate climate systems.

As defensible as these objections once were, today they are obsolete. For one thing, there is now little doubt that, in their gross and aggregate outcomes, the computer models are correct. Moreover, the delayed onset of detectable warming was explained by an important discovery that was made a couple of years ago. Scientists had already known that atmospheric warming is delayed because the surface waters of the oceans store heat at deeper levels before subsequently releasing it. But not until the beginning of this decade did they discover that the same fossil fuel emissions that contribute to global warming by emitting carbon dioxide simultaneously mask that warming by emitting sulfate particles as well. While the carbon dioxide high in the upper atmosphere acts to trap heat inside the global greenhouse, the lower-level umbrellas of sulfate particulates reflect the sunlight back into space, creating localized cooling effects that conceal the continuing warming. "If everyone in the world could magically [remove the sulfates from coal and oil], you would see the fingerprints of global warming in a very short time," says Harvard's McCarthy.

Unfortunately, the sulfate aerosols cannot be considered a long-term neutralizing agent against global warming. For one thing, they remain airborne for only several weeks and mostly in localized areas, while carbon dioxide remains in the atmosphere for one or two centuries. Since both by-products of fossil fuel burning are released simultaneously, the much-longer-lived carbon dioxide will eventually overwhelm the transient sulfates. The sulfates emitted by power plants, factories, automobiles, and volcanoes are regarded by scientists less as an offset than as a mask.

But there is a second reason not to regard sulfate aerosols as a remedy: They are a serious and persistent public and environmental health hazard. They cause low-level air pollution, which visits on us lung disease, crop destruction, and acid rain. A cure of such proven toxicity can scarcely be regarded as a cure at all. At bottom, sulfates are simply another airborne industrial poison generated by the combustion of fossil fuels.

A massive but short-lived pulse of sulfates masked the progression of atmospheric warming when Mount Pinatubo in the Philippines erupted in 1991. But since mid-1992, when the volcanic sulfate debris began to settle out, the planet's surface air temperature has increased by a half-degree Celsius—as much as it had warmed between 1900 and 1990, according to scientists from NASA's Goddard Institute for Space Studies and the National Oceanic and Atmospheric Administration's (NOAA) National Centers for Environmental Prediction.

Earlier generations of computer models were inadequate because their results did not always correspond to the known recorded climate history of the last century. But in the more recent models sulfate aerosols have been included as a significant factor. These models are all still too crude to forecast particular impacts in specific geographical locations—their resolution needs sharpening. But in large gauge they do match the historical record—and the types of impacts they project, I believe, are already beginning to happen.

By mid-June 1995, enormous fires burning in Canadian northern forests had spread into central Canada, expanding at a rate of about 240,000 acres a day. That year fires in Canada's boreal forests consumed more than 3 million acres, an area half the size of the Netherlands. A study by Canadian Forest Service scientists concluded that the northern forest has lost almost a fifth of its biomass over the last 20 years because of enormous increases in fires and insect outbreaks. Before 1970 the forest had absorbed 118 million tons of carbon each year, according to the study—more than counterbalancing Canadian fossil fuel emissions. But in the last decade that balance has shifted, and the forest has released on average 57 million tons of carbon each year. The Canadian Forest Service concluded that the reversal of the forests—from an absorber to an emitter of carbon dioxide—may be contributing to an accelerating buildup of greenhouse gases in the earth's atmosphere.

Nor were the weather extremes confined to northern latitudes. While western portions of Australia experienced record rainfalls in the summer of 1995, the city of Sydney, in eastern Australia, recorded its first rainless August in history.

Several months later, the island of St. Thomas was blasted to shambles by one of fifteen named hurricanes that ravaged the Caribbean that fall. Those devastating storms made 1995 the worst hurricane year since 1933.

In late 1995, after scientists had successfully integrated the cooling effects of sulfates into their climate models, the 2,500-member IPCC reported that it had detected the "fingerprint" of human activity as a contributor to the warming of the atmosphere. "A pattern of climatic response to human activities is identifiable in the climatological record," the IPCC found. In other words, global warming could no longer be attributed to natural climate variability. Furthermore, the scientists observed that the projected warming is due to the increasing quantities of carbon dioxide released by our burning of coal and oil.

The IPCC report was the scientific equivalent of a smoking gun. It should have laid to rest nearly a decade of squabbling over the verification of global warming.

James McCarthy commands a broad overview of the current research. Between 1987 and 1993 he chaired the Scientific Committee for the International Biosphere-Geosphere Programme, a research complement to the IPCC. There he oversaw the work of the leading climate scientists from sixty nations. Dr. Bert Bolin, the Swedish physicist who, until mid-1996, headed the IPCC, served as his vice-chair. In 1996 McCarthy became chair of the Advisory Committee on the Environment of the International Committee of Scientific Unions.

"There is no debate among any statured scientists of what is happening," says McCarthy. By "statured" scientists he means those who are currently engaged in relevant research and whose work has been published in the refereed scientific journals. "The only debate is the rate at which it's happening."

McCarthy echoes concerns that popular accounts of global warming have created a misconception in the public mind. "Questions about global warming—when will it happen, what is the threshold—are the wrong questions," he says. "The real question involves the instability of the climate system. If the world became, on average, ten degrees warmer in the winter and ten degrees cooler in the summer, the average global temperature would remain the same. But the economic and agricultural and ecological effects would be disastrous. The real question is how the planet will be affected by the extremes of a climatic instability that could put whole segments of, say, agriculture out of business at either end of its spectrum of extremes."

Noting that the recent phenomenon of five consecutive El Niño years (marked by a vast pool of warm water in the eastern equatorial Pacific Ocean) is unprecedented, McCarthy wonders whether "we have already tripped into a new climate regime where the old anomalies have become the new norm. And we have yet to appreciate the biological impacts of prolonged warming. The concern of many scientists is that by pushing the concentrations of atmospheric carbon dioxide to higher and higher levels, we are directing the system into a new state that may be considerably less stable than what we have enjoyed for the past several hundred thousand years."

El Niño is associated with intense storms in the Pacific Ocean, and it promotes severe droughts in continental interiors. Several scientists note that the current, extended El Niño may be followed by its opposite—a protracted La Niña, whose upwelling of cold waters from the ocean depths could actually cool the climate for the next several years. While that would add even more confusion to the public perception of global warming, it would nevertheless be consistent with the intensifying weather oscillations that mark the erratic and nonlinear processes of climate change.

In November 1995, following record rainfalls and extensive flooding in France and the Netherlands, Typhoon Angela—the most powerful storm in nearly two decades—killed more than 600 Filipino residents.

At the end of the year frustrated officials were forced to cancel the World Cup ski tournament in Austria because record high temperatures and a lack of snow cover made skiing impossible.

At the same time northern portions of Mexico experienced the coldest winter in 25 years, and on New Year's Day 1996, Mexico City saw its first snowfall in 20 years.

That same month, when Sapporo, Japan was buried by unprecedented levels of snow, officials had to call in the military to help dig out the city.

A second smoking gun—proof of the increasing instability of the world's climate—was discovered in 1995, when a team of scientists at the NOAA's National Climatic Data Center verified an increase of extreme weather events in the United States, the former Soviet Union, and China (the only areas for which they examined the data). These areas, they found, are experiencing more extreme rain and snowfalls, more winter precipitation, more droughts and floods, and more extremely warm weather. The team of scientists, headed by Dr. Thomas Karl, compiled an index that showed elevated minimum nighttime and winter temperatures, higher-than-normal levels of precipitation during winter months, a higher-than-normal number of extreme one-day rain or snowfalls in the United States, and unusually severe droughts during summer.

The NOAA team's findings are extremely significant. The signature of the recent climate record, they found, is unlike natural variability. Rather, it resembles the kinds of changes that would be expected from escalating emissions generated by the burning of oil and coal. The scientists concluded that the growing weather extremes are due, by a probability of 90 percent, to rising levels of greenhouse gases. Specifically, they declared that the climate in the United States is becoming more "greenhouse-like." In scientific terms, they concluded that "the late-century changes recorded in U.S. climate are consistent with the general trends anticipated from a greenhouse-enhanced atmosphere."

The events they identified, in other words, are precisely what the corrected climate models project.

In February 1996 Texas residents shivered through ice, snow, and single-digit temperatures—the coldest weather to hit the region in almost a decade. But an extraordinary March brought with it record temperatures—in the mid-nineties. As a result, northwest of Forth Worth, grass and brush fires blackened a swath of land the size of Manhattan. The fires continued intermittently for a month, scorching more than 300,000 acres in one of the worst fire seasons in Texas history.

In March 1996 a deadly blizzard in the western Chinese highlands prompted appeals for international aid to avert fam-

ine. At least 60,000 ethnic Tibetan herders in Qinghai province and Tibet faced starvation from storms that drastically reduced their food supply, wiping out 750,000 head of livestock. The storms, accompanied by record-setting lows of minus 49 degrees, killed 48 herders in Sichuan, according to reports. According to Chinese officials, "This year's snowfall is four times greater than last year."

By March 1996, more than 20 percent of Laos's rice paddies had been decimated by five successive years of floods, droughts, and pest attacks, according to reports from Reuters. UN officials estimate the country faces a shortage of 132,577 tons of rice—which amounts to 75 percent of the caloric intake of the Laotian people. One official estimated that without immediate food aid, nearly 10 percent of Laotians are at risk for malnutrition and starvation. The official noted that the country's persistent and worsening food crisis was propelled into a full-fledged disaster by last year's floods—the worst in Laos in 30 years.

The most striking aspect of the recent severe weather events is that virtually all of them set a new record. Are these events absolute proof of global warming? Not taken alone—anecdotal evidence never is. But in combination with the gathering weight of other evidence—both from scientific laboratories and from the world's oceans, glaciers, and forests—they present an urgent and compelling case. Nor was 1995 an aberration. The severe weather has continued into 1996. My own backyard became a snow-buried casualty of New England's 1995–96 winter from hell. As late as May 13, we experienced record low temperatures. Eight days later, the temperature set a new record high, for that date.

The naysayers have dubbed global warming "the mother of all environmental scares." Despite its sarcasm, that dismissal contains a recognition that potentially destabilizing social and economic and philosophical issues cluster around the phenomenon of climate change.

The threat of climate change challenges the capacities of our governments and imperils the basic processes of democracy. And it highlights with startling ferocity a basic shortcoming of our economic system: That system ignores the fact that the global economy is rooted in and bounded by the global environment. Our system of economic accounting counts as profit all the coal and oil the world consumes, but it fails to enter the resulting environmental damage anywhere on the global balance sheet.

It is not only the oil and coal spokesmen who deny the evidence of climate change. A number of economists, political commentators, and conservative ideologues downplay the urgency of the issue as well. Some say that if we simply allow the free market to flourish over the next several decades, we will accumulate enough capital to pay for remedial actions. Huge economic dislocations would be involved in switching to new sources of energy, they argue. Paying to remedy the damage later, they contend, will be no more costly and less disruptive than paying for preventive measures now.

Several years ago James Watkins, then secretary of energy in the Bush administration, expressed horror at the prospect of making an economic shift as extensive as global warming demands. "What if the science is wrong?" he asked me. "Can you imagine how much it would cost?" "What if it's right?" I responded. "What would the costs be then?" Watkins did not respond.

The "pay later" argument is absurd on its face. Numerous economists have shown that the longer we delay making an energy transition from fossil fuels to renewable energy, the costlier that transition will be, as the pressures of environmental deterioration reduce the time we have to plan and implement the transition.

Other naysayers point to estimates that certain agricultural areas may actually benefit from a slight warming and subsequent longer growing seasons. A little bit of warming isn't really so bad, they say, ignoring the fact that crops in the developing world would be decimated. But they fail to acknowledge how intricately the earth's climatic and ecological systems are interrelated. Scientists use the term *feedbacks* to refer to the complex and often unanticipated chains of events that can ripple through the ecosphere when one part of it is perturbed.

Take baby seals, for example.

Several springs ago, a large number of seal pups were found to be starving to death on the beaches and rocks of the northern California coast. Responding to the alarm, hordes of well-meaning volunteers—armed with baby bottles and diced fishing bait—converged on the area and eventually nursed many of the pups back to health. But the starving pups—like the proverbial canary in the coal mine—were an expression of a larger problem. Every spring seal pups feed on fish that inhabit the surface waters off northern California. That spring, however, the surface waters had warmed so much that the fish descended to colder, deeper levels to survive. Unfortunately, their new habitat was deeper than the seal pups could dive—leaving them stranded and malnourished until the sympathetic humans rescued them.

Some feedbacks may be induced by even a slight increase in average global temperatures. The most powerful "greenhouse gas"—and the one most responsible for global warming—is atmospheric water vapor. At a concentration thirty times greater than that of carbon dioxide, atmospheric water vapor cannot be directly affected by increasing or reducing oil and coal emissions, according to Dr. Michael Oppenheimer, a former researcher at the Harvard-Smithsonian Center for Astrophysics and now a senior scientist for the Environmental Defense Fund. Rather, its effect is indirect: Even the slight warming caused by the carbon dioxide buildup leads to more evaporation of ocean waters, which in turn causes a disproportionate increase in the amount of heat-trapping water vapor in the atmosphere.

A more dramatic feedback could accelerate the thawing of the Alaskan tundra. For thousands of years on the northern slope of Alaska, the plants of the tundra have been taking up carbon dioxide. When these plants die, before they can decay, they are frozen in deep layers of permafrost, where they store carbon dioxide. But in 1993 scientists discovered that this process has reversed itself and that the tundra has begun to thaw, releasing stored carbon dioxide into the atmosphere.

Dr. Walter Oechel, of San Diego State University, attributes this unanticipated thaw, which began sometime before 1981, to a marked warming trend in the high Arctic, where the mean summer temperature rose from 35 degrees Fahrenheit in 1971 to 41 degrees twenty years later. If the global temperature increases even slightly, it could accelerate the release of huge amounts of CO_2 (as well as methane, another potent greenhouse gas), which in turn could hasten the retreat of the Arctic ice pack, which could trigger a cascade of uncontrollable catastrophic reactions. The Arctic ice cover, at present, cools the ocean currents and reflects sunlight back into space. But when that ice cover melts and exposes the dark soil underneath, the Arctic landmass will absorb—rather than reflect—sunlight, thereby accelerating the warming of the oceans.

What most frightens scientists is this possibility of runaway feedbacks. To be sure, global warming is not likely to be a smooth, gradual, linear process. But scientists do not know at what threshold a slight increase in warming—accelerated by feedback effects—could abruptly catapult the planet into an entirely new climate regime.

The skeptics who say a little warming isn't so bad are betraying not only our scientific knowledge of natural systems but our first-hand recognition of their exquisitely complex intricacies. Their view insults our basic sense of reality.

If all our scientific endeavors—to understand the physical universe, the human body, and the planetary environment—share one lesson, it is this: Throughout human history, the complexities of nature have consistently surprised and outflanked the human mind. Did the last generation of pharmaceutical researchers anticipate that microbes would resist the cures they invented, generating new forms of immunity and, in some cases, increased virulence? Did their pharmaceutical forefathers foresee the side effects of the cures they invented? The casualties are in the literature.

Regardless of the pronouncements of climate scientists on both sides of the warming argument, this fact is inescapable: We are, through our industrial activities, tampering with immense planetary systems whose complexities and interactions we barely understand.

"The inhabitants of planet Earth are quietly conducting a gigantic environmental experiment," says Dr. Wallace S. Broecker of Columbia University's Lamont-Doherty Earth Observatory. "So vast and sweeping will the consequences be that, were it brought before any responsible council for approval, it would be firmly rejected."

A third important contribution to our understanding of the global climate appeared in the spring of 1995, when David J. Thomson, a signals analyst at AT&T Bell Labs, published his evaluation of a century of summer and winter temperature data. While some greenhouse skeptics had attributed this century's atmospheric warming to solar variations, Thomson discovered the opposite: The accumulation of greenhouse gases had overwhelmed the relatively weak effects of solar cycles on the climate. But Thomson also discovered something that may explain the unseasonal nature of some of today's extreme weather events. Around the beginning of World War II, accelerating industrialization led to a skyrocketing of carbon dioxide emissions. At that point, Thomson found, the timing of the seasons began to shift. After 1940, he wrote in the journal *Science,* the seasonal patterns "of the previous 300 years began to change and now appear to be changing at an unprecedented rate."

In the northern hemisphere spring is now arriving a week earlier than it did twenty years ago because of increased concentrations of carbon dioxide, according to Drs. Charles Keeling and T. P. Whorf of the Scripps Institution of Oceanography and J. F. S. Chin of the Mauna Loa Observatory in Hawaii. These researchers wrote in 1996 that the changes in seasonal timing "reflect increasing assimilation of carbon dioxide by land plants in response to climate changes accompanying recent rapid increases in temperature."

This point is important. One of the major arguments of the naysayers has been to contend that the planet did not warm significantly between 1940—when emissions of greenhouse gases began to skyrocket with the wartime industrial mobilization—and the mid-1970s, when temperatures took an upward course. Between 1940 and 1970, much of that newly generated carbon dioxide may have been absorbed by forests and oceans, and its effects may have been temporarily neutralized by the sulfate cover. But it is precisely since 1940 that the timing of the seasons has changed. This change could be a significant driving force behind some of the recent extreme weather events.

The worst possible consequences of global warming involve more than an alteration in the seasons or even a steady increase in global temperatures. The scientists' most cataclysmic, if improbable, scenario is based on paleontological records culled from ocean sediments and ice-core samples from glaciers. Until recently, scientists believed the transitions between ice ages and more moderate climatic periods had occurred gradually, over several centuries.

No longer.

Ancient ice cores are made of annual layers of frozen water, which hold natural records of our ancient atmosphere. Several years ago researchers examining them found that those ice age transitions, involving temperature changes of up to 10 degrees Celsius, occurred within the space of only ten years—a virtual millisecond in geological time. In the last 70,000 years, they learned, the earth's climate has snapped—abruptly and dramatically—into radically different temperature regimes. "Our results suggest that the present climate system is very delicately poised," said Scott Lehman, a researcher at the Woods Hole Oceanographic Institution, announcing findings in 1993. "The system could snap suddenly between very different conditions with an abruptness that is scary. It's a strongly non-linear response, meaning shifts could happen very rapidly if conditions are right, and we cannot predict when that will occur. Our studies tell us only that when a shift occurs, it could be very sudden." In an interview with *The Boston Globe* he added, "You don't want to push your luck by perturbing the system. A small effect might produce a major change."

Lehman's cautionary tone is underscored by findings that the end of the last ice age, some 10,000 years ago, was marked by a series of extreme oscillations between warming spikes and severe regional deep freezes. As the surface waters of the North Atlantic warmed, the Woods Hole team found, rising temperatures triggered snowmelts in the Arctic and increased rainfall in the northern latitudes. That infusion of fresh water diluted the salt content of the ocean, which, in turn, changed the course of the deep-ocean warming current from a northeasterly direction to one that ran nearly due east. Should such an episode occur today, the researchers concluded, "the present climate of Britain and Norway would change suddenly to that of Greenland and Northern Canada."

The foremost researcher of deep-ocean-current dynamics, Wallace Broecker of Lamont-Doherty, observed in November 1995 that an ocean conveyor current "shutdown or comparable drastic change is unlikely, but were it to occur, the impact would be catastrophic. The likelihood of such an event will be highest between 50 and 150 years from now, at a time when the world will be bulging with people threatened by hunger and disease and struggling to maintain wildlife under escalating environmental pressure. It behooves us to take this possibility seriously. We should spare no effort in the attempt to understand better the chaotic behavior of the global climatic system."

The naysayers' rallying cry of last resort is uncertainty. We know too little about climate change to act, they assure us. Until the holes of scientific uncertainty are filled, they warn, it would be irresponsible to act—especially when action could be costly and, worse, so revolutionary as to disrupt the established order of things.

What they do not mention is that to avoid acting could be to compound, incalculably, the costs of addressing climate change and its disruptions to civilization. What they do not mention is that uncertainty cuts both ways.

At a recent meeting at Tufts University, Dr. Richard Lindzen of the Massachusetts Institute of Technology, arguably the most academically accomplished of the scientific skeptics, described at great length the various shortcomings of the climate models and their inability to resolve a number of significant uncertainties. When he had finished, Dr. Michael McElroy, chairman of Harvard University's department of earth and planetary sciences, recalled that in the early 1980s scientists had spent several years modeling projected ozone depletion. "When researchers finally conducted actual ozone measurements in the atmosphere, their findings were far worse than the worst case scenarios of the models," he said, adding, "Just because a situation is uncertain does not imply that the underlying reality is benign."

Our scientific knowledge, in other words may even be lagging behind nature. The momentum of globally disrupting climate change may be further advanced than earth science, with its areas of uncertainty, is currently able to prove.

What we do know is that the earth's systems are showing irrefutable signs of climate-related stress. The evidence goes beyond computer models and laboratory calculations. It lies in numerous research discoveries (examined later in this book) about the oceans, the forests, the glaciers, and the soils, and in the dramatic outbreaks of infectious diseases under the forces of climatic change.

Moments like these are what we make of them. They can swamp us with a paralyzing hopelessness, or they can inflame us with a new sense of purpose. What is needed, I think, is the social counterpart to a climate snap—a rapid, immense, worldwide gathering of political will.

When I was young, my father told me the countries of the world would make peace only when they were threatened by invaders from outer space. Today it is climate change that poses a common threat to all of humanity. Yet the situation also contains a hopeful potential within its grim prospect—a moment of extraordinary, thrilling opportunity.

Under all the political instability that marks the end of this millennium lurks the single message that the world we have created needs to be changed. Too many elements of our social and economic systems, far from serving human needs, are now frustrating our survival as a civilized species.

Finding solutions—given the fury with which the battle over climate change is being waged—will require a massive mobilization of our determination. It will mean putting aside, at least temporarily, many things that divide us. It will demand of us a huge leap in thinking—and mustering the collective will to force the changes our fevered atmosphere requires.

The resistance will be that strong. Necessity for change and the requirements for our survival are that great.

TWO

The Battle for Control of Reality

THE FINANCIAL RESOURCES AVAILABLE TO THE OIL AND coal lobbies are almost without limit. They can buy Congress. In fact, long before the climate issue surfaced, they already had.

They can buy media access. Not just the Mobil ads, eye-catchingly conspicuous on the *New York Times* op-ed page, but also access to editorial boards, TV producers, and every relevant reporter in the country.

Over the last six years the coal and oil industries have spent millions of dollars to wage a propaganda campaign to downplay the threat of climate change. Much of that money has gone to amplifying the views of about a half-dozen dissenting researchers, giving them a platform and a level of credibility in the public arena that is grossly out of proportion to their influence in the scientific community.

The campaign to keep the climate change off the public agenda involves more than the undisclosed funding of these "greenhouse skeptics." In their efforts to challenge the consensus scientific view about the escalating turmoil of the global atmosphere, the public relations apparatus of the oil and coal industries has publicized weather reports, and funded and distributed books and videos and self-proclaimed journals of science that are dismissed by the vast majority of mainstream scientists.

John Stauber is the editor of *PR Watch,* a newsletter that monitors the $10-billion-a-year public relations industry. In his book *Toxic Sludge is Good for You,* he quotes one public relations executive as saying: "Persuasion by its definition is subtle. The best PR ends up looking like news. You never know when a PR agency is being effective. You'll just find your views slowly shifting." The quote applies, with uncanny accuracy, to the campaign to carefully manufacture public confusion about global climate change. Big oil and big coal have successfully created the general perception that climate scientists are sharply divided over the extent and the likely impacts of climate change—and even over whether it is taking place at all.

The Information Council on the Environment (ICE) was the creation of a group of utility and coal companies. In 1991, using the ICE, the coal industry launched a blatantly misleading campaign on climate change that had been designed by a public relations firm. This public relations firm clearly stated that the aim of the campaign was to "reposition global warming as theory rather than fact." Its plan specified that three of the so-called greenhouse skeptics—Robert Balling, Pat Michaels, and S. Fred Singer—should be placed in broadcast appearances, op-ed pages, and newspaper interviews.

With all the sophistication of modern marketing techniques, the ICE campaign targeted "older, less-educated men" and "young, low-income women" in electoral districts that get their electricity from coal and that, preferably, have a congressperson on the House Energy Committee. The campaign was clever if not accurate. One newspaper advertisement prepared by the ICE, for example, was headlined: "If the earth is getting warmer, why is Minneapolis getting colder?" (Data indicate that the Minneapolis area has actually warmed between 1 and 1.5 degrees Celsius in the last century.)

The ICE campaign had barely gathered momentum when it was exposed by environmentalists, who provided information about it to the media. Several embarrassing news articles led to the ICE's demise. But its case illustrates how a little "repositioning" can generate tremendous public confusion about the state of scientific understanding.

In 1995 I co-authored an article for *The Washington Post* on new research that linked disease outbreaks in various places around the world to changes in local climate. The article was generally well received, but a number of readers wrote to assure me that there was no proven link between industrial emissions and global climate change. Their responses caused me to wonder why so many people believed that the validity of the issue was still so far from certain.

It did not take me long to find the answer.

Ever since climate change took center stage at the 1992 UN Conference on Environment and Development (UNCED) in Rio de Janeiro, Pat Michaels and Robert Balling, together with Sherwood Idso, S. Fred Singer, Richard S. Lindzen, and a few other high-profile greenhouse skeptics have proven extraordinarily adept at draining the issue of all sense of crisis. They have made frequent pronouncements on radio and television programs, including a number of appearances by some of them on the Rush Limbaugh show; their interviews, columns, and letters have appeared in newspapers ranging from local weeklies to *The Washington Post* and *The Wall Street Journal*. In the process they have helped create a broad public belief that the question of climate change is hopelessly mired in unknowns.

If the climate skeptics have succeeded in confusing the general public, their influence on decision makers has been, if anything, even more effective.

Their testimony contributed to the defeat of proposals in California and Colorado to increase electricity rates to reduce the amount of greenhouse gases produced by oil- and coal-burning power plants. (A similar initiative was approved recently by the state of Minnesota.) Congressional conservatives have used the testimony of the skeptics to justify cutting the climate research budgets and to discredit the scientific findings of the IPCC.

The origins of the prominence of most of these greenhouse skeptics are spelled out, remarkably enough in several annual reports of the $400 million coal giant, the Western Fuels Association. In its 1994 annual report, Western Fuels declared quite candidly that "there has been a close to universal impulse in the [fossil fuel] trade association community here in Washington to concede the scientific premise of global warming . . . while arguing over policy prescriptions that would be the least disruptive to our economy. . . . We have disagreed, and do disagree, with this strategy."

Western Fuels elaborated on its approach in another report: "When [the climate change] controversy first erupted at the peak of summer in 1988, Western Fuels Association decided it was important to take a stand. . . . [S]cientists were found who are skeptical about much of what seemed generally accepted about the potential for climate change. Among them were [Pat] Michaels, Robert Balling of Arizona State University, and S. Fred Singer of the University of Virginia. . . . Western Fuels approached Pat Michaels about writing a quarterly publication designed to provide its readers with critical insight concerning the global climatic change and greenhouse effect controversy. . . . Western Fuels agreed to finance publication and distribution of *World Climate Review* magazine."

In 1991, before the ill-fated ICE campaign had been buried, one of its funders, Western Fuels, spent $250,000 to produce a video. The video was shown extensively in Washington as well as in the capitals of the OPEC nations. Insiders at the Bush White House said it was Chief of Staff John Sununu's favorite movie—he showed it that often. Then-Secretary of Energy James Watkins cited it in a conversation during a visit to *The Boston Globe,* where I interviewed him. The video's aim is to persuade policymakers that a warmer, wetter, carbon dioxide–enhanced world would be, contrary to the alarms of environmentalists, a godsend.

Titled *The Greening of Planet Earth,* the video is narrated by Sherwood Idso, an active skeptic, and it features Richard Lindzen as well as a number of botanists and agronomists. (The video was produced by a company headed by Idso's wife; the coal industry, which funded it, has purchased and circulated hundreds of copies of books and publications written by Idso.) In near-evangelical tones, it promises that a new age of agricultural abundance will result from the doubling of the atmospheric concentration of carbon dioxide. It shows plant biologists predicting that yields of soybeans, cotton, wheat, and other crops will increase by 30 to 60 percent—enough to feed and clothe the earth's expanding human population. The video portrays a world where vast areas of desert are replaced by grasslands, where today's grass- and scrublands are transformed by a new cover of bushes and trees, and where today's diminishing forests are replenished by new growth as a result of a nurturing atmosphere of enhanced carbon dioxide.

Unfortunately, it overlooks the bugs.

Insects are extremely sensitive to changes in temperature. According to a panel of the World Health and World Meteorological organizations and the UN Environmental Programme (UNEP), even a minor elevation in temperature would trigger an explosion in the planet's insect population. A slight warming, the panel suggests, could result in insect-related crop damage, leading to a significant disruption in the food supply. The spread of insect-borne diseases could surge. The panel, which examines interactions between global climate and biological systems, notes that the *Aedes aegypti* mosquito, which spreads dengue fever and yellow fever, has traditionally been unable to survive at altitudes higher than 1,000 meters because of colder temperatures there. But with recent warming trends, those mosquitoes have now been reported at 1,240 meters in Costa Rica and at 2,200 meters in Colombia. Malaria-bearing mosquitoes, too, have moved to higher elevations in central Africa, Asia, and parts of Latin America, triggering new outbreaks of the disease.

It appears that while the coal-funded video extolled the benefits of global warming, it neglected to tell people what the warming-induced infestation of termites in New Orleans may be trying to tell them now.

(In fairness, many agricultural scientists also acknowledge some short-term benefits of enhanced CO_2. In the short term, it may indeed increase yields and growth rates of food crops in the mid-northern latitudes—to the benefit of U.S., Canadian, and Russian agriculture. But other effects are not so positive. The initial increases in crop growth and food yield, many scientists fear, will soon flatten, and a long-term diet of concentrated carbon dioxide will weaken the plants, making them less robust.

(More importantly, enhanced CO_2 would be devastating to crop growth in the poor areas of the world—the midtropical regions. In many of these areas, population growth is already stressing the food-growing capability of the land. Increased CO_2, agricultural researchers say, will force more rapid plant respiration. When that forcing is accompanied by slightly elevated temperatures, the plants will stop growing and their yield will dwindle. A significant enhancement of CO_2 in the earth's most heavily populated middle-warm belt will accelerate rates of malnutrition, disease, and starvation.)

Western Fuels is not alone in its efforts to veil the reality of climate change. On the eve of the March 1995 round of international climate negotiations (which had been set up by the UNCED conference in Rio), another industry lobbying group, the Global Climate Coalition (GCC), disseminated a report issued by the private weather-forecasting firm Accu-Weather. The report declared there has been no increase in severe weather events. A GCC press release that accompanied the report noted that "Accu-Weather experts say there is no convincing evidence that global weather is becoming more extreme." It quoted an Accu-Weather executive as saying, "Scientific evidence squarely disputes the hypothesis that hurricanes are becoming stronger and more frequent, that tornadoes have increased in number, and that droughts and floods are becoming more common. In fact, the data show that . . . temperature and precipitation extremes are no more common now than they were 50 to 100 years ago."

The report was dismissed by a number of mainstream scientists, who noted that it contradicted the findings of a team of researchers from the NOAA National Climatic Data Center earlier that year. The NOAA team had demonstrated that the increase in severe weather events had been fueled by atmospheric warming. Several scientists took issue with the Accu-Weather report's methodology, pointing out that its weather and temperature readings had been taken from only three data points—Augusta, Georgia, State College, Pennsylvania, and Des Moines, Iowa—hardly a broad-based sample.

Equally telling, the Accu-Weather report flies in the face of insurance industry figures that show that annual weather-related disaster claims have increased sixfold since the 1980s, from $5 billion to $30 billion in the first half of the 1990s.

In May 1995 Judge Allan Klein, who sits on the Minnesota Public Utilities Commission, was charged with the responsibility of determining the environmental costs of the burning of coal by Minnesota power plants. In his administrative courtroom in St. Paul, Judge Klein heard testimony from four greenhouse skeptics who had been hired as expert witnesses not only on behalf of Western Fuels Association, but of several local utilities and the state of North Dakota, the largest supplier of coal to neighboring Minnesota.

Called to the stand was Richard Lindzen, a professor of meteorology at MIT, who testified that, given current trends, the likeliest increase in atmospheric warming by the middle of the next century would be 0.3 degrees Celsius. Although global emissions of carbon dioxide and other greenhouse gases will actually double by the year 2040, Lindzen, a short, owlish-looking man with a professorial demeanor, an argumentative style, a quick sense of humor, and $2 million in federal research grants in his distinguished thirty-year academic career, believes the impacts, if any, will be negligible.

Also called to testify was Pat Michaels, associate professor of climatology at the University of Virginia, who told Judge Klein that despite the buildup of greenhouse gases, he foresees no increase in the rate of sea level rise—a feared consequence of global warming. The gregarious, engaging Michaels is a frequent commentator on climate issues and the founder and publisher of *World Climate Review* and its successor publication, *World Climate Report.*

Robert Balling, a professor of climatology at Arizona State University, concurred that any potential warming is barely worth consideration. "I would anticipate no more than a small rise in temperature, maybe a degree," he testified in St. Paul. "Whether it will continue to move linearly or continue to move in some other fashion is something that's so speculative, it's almost useless to think about it." Balling, a Marine-trim, boyish-looking 43-year-old author of a book on global warming titled *The Heated Debate,* has labeled concerns about climate change "pure media hype."

The power of this apparent certainty should not be underestimated.

Professor Willett Kempton, a senior policy scientist at the University of Delaware, has documented the influence of the greenhouse skeptics in *Environmental Values in American Culture,* a book he coauthored that was funded by the National Science Foundation. An aide to a Republican congressman who supports health-related environmental regulation and endangered species protection told Kempton about a televised presentation he had heard by one skeptic. "It came out pretty clearly . . . that there is a range of disagreement, and [that] two people with equally impressive credentials can disagree," he said. After hearing several more skeptics, the aide decided that scientifically "there's no mainstream, there's no fringes, there are just people all over the lot." Moreover, a union lobbyist interviewed for the same book said: "Everything stems from the assumption that the earth is getting warmer and the causes are [greenhouse gases]. . . . But when I read opposing articles . . . they're actually . . . more persuasive than the others." The skeptics led him to believe that "very little [has been] done to measure changes in temperature of oceans or the air mass above the oceans. And there has not really been the depth of scientific inquiry necessary to say that . . . is a problem."

The tiny group of dissenting scientists have been given prominent public visibility and congressional influence out of all proportion to their standing in the scientific community on the issue of global warming. They have used this platform to pound widely amplified drumbeats of doubt about climate change. These doubts are repeated in virtually every climate-related story in every newspaper and every TV and radio news outlet in the country.

By keeping the discussion focused on whether there really is a problem, these dozen or so dissidents—contradicting the consensus view held by 2,500 of the world's top climate scientists—have until now prevented discussion about how to address the problem.

The skeptics are virtually unanimous in accusing their mainstream scientific colleagues of exaggerating the magnitude of the climate problem in order to perpetuate their own government research funding.

But that argument cuts both ways.

While testifying in St. Paul, Pat Michaels revealed under oath that he had received more than $165,000 in industry and private funding over the previous five years—funding he had never previously disclosed. Not only did Western Fuels find both his publications, he disclosed, but it provided a $63,000 grant for his research. Another $49,000 came to Michaels from the German Coal Mining Association. A smaller grant of $15,000 came from the Edison Electric Institute. Michaels also listed a grant of $40,000 from the western mining company Cyprus Minerals. Questioned by the assistant attorney general about that grant, Michaels responded, "You know, with all due respect, you're going to think I'm not telling the truth. I'm trying to remember directly what came out of the project. . . . I'm sure we were looking at regional temperatures in some way."

In fact, Cyprus Minerals was, at the time, the largest single funder of the virulently antienvironmentalist Wise Use movement. The biggest organizational member of that movement was a group called People for the West!, whose largest funder, with at least $100,000 in donations, was Cyprus Minerals. According to the Clearinghouse in Environmental Advocacy and Research, as recently as 1995 Cyprus Minerals' director of governmental affairs was a member of the board of directors of People for the West!.

In interviews, Michaels has insisted that he dissociated himself from the ICE campaign when he learned of what he called its "blatant dishonesty." But he apparently had no qualms about accepting money to publish his own journal, *World Climate Review,* from one of the same coal industry sources that funded the ICE campaign. (The industry funding of Michaels's publications was first made public by Bud Ward, editor of *Environment Writer,* the newsletter for journalists published by the Environmental Health Center of the National Safety Council. Unfortunately the journalists Ward writes for made little use of the information.) Michaels, for his part, insists that this now-defunct journal, as well as its successor coal-funded publication, *World Climate Report,* which Michaels also edits, are serious journals of climate science.

However, a reading of those publications reveals passages such as this one, written by Michaels in the fall 1994 issue of *World*

Climate Review: "The fact is that the artifice of climate-change-as-apocalypse is crumbling faster than Cuba. . . . There is genuine fear in the environmental community about this one, for the decline and fall of such a prominent issue is sure to horribly maim the credibility of the green movement that espoused it so cheerily."

This is not the language of science, such as one finds in *Science, Nature,* or *The Bulletin of the American Meteorological Society.* It is the language of propaganda.

The winter 1993 issue of Michaels's magazine featured a cover photo that appeared to replicate the front page of *The Washington Post,* with the headline "The End Nears Again." Michaels was subsequently forced to apologize to the *Post* for his choice of cover art, which more closely resembled a cover for the *National Lampoon* than one for a journal of science. In the winter 1993 issue, he wrote of mainstream scientists in words that would be devastating if they were applied to his own career and its sponsors: "The fact is that virtually every successful academic scientist is a ward of the federal government. One cannot do the research necessary to publish enough to be awarded tenure without appealing to one or another agency for considerable financial support. . . . Yet these and other agencies have their own political agendas."

A critical point that Michaels chooses to ignore is that all research sponsored by the federal government is subjected to the exacting requirements of scientific proof. In what is called the "refereed" literature, one's research peers systematically review an article as a condition of publication. By contrast, private, industry-funded research is not necessarily peer reviewed and is frequently published in industry journals without undergoing this kind of rigorous scientific scrutiny.

But Michaels's argument founders on a far more obvious rock: If federally funded scientific research is merely a conspiracy to milk the public, why does the government permit it? And why, of all things, would it want to encourage reports of climate change? The federal government is already facing far more problems today than it has resources to handle. What conceivable reason would it have for funding scientists to exaggerate evidence of a coming climate catastrophe? Last time I looked, Congress had not approved any funding for agencies to discover remote, highly implausible new crises that would require greater public expenditures. Although Michaels has said he supports continued federal research funding, his testimony was cited by the chairman of the House Subcommittee on Energy and the Environment as a basis for cutting programs critical to monitoring climate change.

Questions of his funding aside, Michaels's statements have frequently blurred the roles of scientist and propagandist for his and his supporters' conservative political views. These views include bashing the United Nations.

In the spring 1996 issue of *World Climate Report,* Michaels reviewed the federal government's 1996 *State of the Climate* document. "If this is an official document," Michaels wrote, "there's no doubting that our federal government is a principal broadcast organ for the views of the United Nations IPCC. . . . It's obvious that the U.N. is viewed by the current administration as the defining entity for our climate."

Michaels's connections were further clarified in an article he authored in a 1993 issue of *World Climate Review.* This extensive article was essentially a retread of the Western Fuels video touting the beneficial effects of carbon dioxide. One source Michaels cited was Sherwood Idso's son, Keith E. Idso, a doctoral candidate at Arizona State University. Keith Idso is another skeptic who was hired by Western Fuels to testify at the St. Paul hearing.

Idso's testimony in St. Paul provided a moment of public embarrassment to his coal sponsors and a touch of comic relief for the audience. On the stand, he was asked about an article he had written titled "The Greening of the Planet." The article, which had appeared in a magazine called the *New American,* detailed in a fairly clinical scientific style his experiments on the effects of enhanced carbon dioxide on sour orange trees. But it concluded with a startling burst of political rhetoric: "This good news [about enhanced carbon dioxide] is not what those intent on destroying our freedoms and imposing their will on the nations of the earth want us to hear, and they skillfully promote alternative voices to confuse the issue. The truth, however, will not be suppressed."

Assistant Attorney General Jonathan Wirtschafter asked Idso

on the witness stand, "Mr. Idso, do you know if the *New American* is published by an advocacy group or a research institute?"

"I know it's not a scientific magazine," Idso replied. "It's something in the popular press."

"Is it published by an advocacy group of some sort?" Wirtschafter asked.

"I don't know if it's advocacy. I know it's some political type organization."

"What organization is that?"

"I can't remember," Idso said. "Some kind of society, I think."

"Was it the John Birch Society?" Wirtschafter asked.

Idso conceded that it was.

What is so extraordinary about the public career of Pat Michaels is that even after his initial association with the extremely cynical coal-funded campaign known as ICE, even after his publication of two journals financed by the coal industry, even after his receipt of money from such flagrantly ideological sources as the largest funder of the Wise Use movement and his use of source material published by the John Birch Society, he has nonetheless appeared as a star witness at several congressional hearings, most notably before the House Science Committee. There Michaels's testimony has been accorded more scientific credibility than that of scientists like Dr. Jerry Mahlman, director of NOAA's Geophysical Fluid Dynamics Laboratory at Princeton University; Dr. Michael MacCracken, a leading climate modeler at Lawrence Livermore National Laboratory for twenty-five years and later director of the largest federal climate science effort, the U.S. Global Change Research Program; and Dr. Robert Watson, co-chair and lead author of the 1995 IPCC report on the impacts and uncertainties of global climate change.

The case of Robert Balling is equally intriguing. A geographer by training, much of Balling's research prior to 1990 focused on hydrology, precipitation, water runoff, and other southwestern water- and soil-related issues. Since 1991, however, the year he was solicited by Western Fuels, Balling has emerged as one of the most visible and prolific of the climate change skeptics.

Beginning with his work for the ICE campaign, Balling has

also received, either alone or with colleagues, nearly $300,000 from coal and oil interests in research funding, which he has never voluntarily disclosed. In his collaborations with Sherwood Idso, Balling has received about $50,000 in research funding from Cyprus Minerals, as well as a separate grant of $4,900 from Kenneth Barr, who at the time was CEO of Cyprus. The German Coal Mining Association has provided about $80,000 in funding for Balling's work. The British Coal Corporation has kicked in another $75,000. Balling disclosed his industry funding under oath during the administrate hearings in Minnesota in 1995.

Given the obvious economic interests of OPEC in the climate debate, it is not surprising that Balling has also received a grant of $48,000 from the Kuwait Foundation for the Advancement of Science, as well as unspecified consulting fees from the Kuwait Institute for Scientific Research.

Balling's 1992 book, *The Heated Debate,* was published by a conservative think tank, the Pacific Research Institute, one of whose goals is the large-scale repeal of environmental regulations. Balling's book was subsequently translated into Arabic and distributed to the governments of the OPEC nations. The funding for this edition of his book was provided by the Kuwait Institute for Scientific Research.

Just because research is funded by industry, to be sure, it is not necessarily tainted. But public disclosure of industry funding is of critical importance so that the research can be reviewed for possible bias. That disclosure requirement is mandatory in other areas of science. If a medical researcher's work is funded by, say, a pharmaceutical company, professional ethics demand that such funding be disclosed in a tagline, when the work is published in the *New England Journal of Medicine* or the *Journal of the American Medical Association.* It is unfortunate that the same standards of scientific and professional ethics do not extend to the field of climate science.

In late 1995 Balling authored an op-ed piece in *The Wall Street Journal* headlined "Keep Cool About Global Warming." Here he attacked the integrity of the IPCC, declaring that the panel's summaries are written "by a few group leaders, and it opens the door for slanting the underlying message of the comprehensive document.

News accounts [based on those summaries] misrepresent reality when they use selective information, offer worst-case scenarios and make claims about increased confidence in the scientific community about predictions of potentially catastrophic climate changes."

It is understandable that a reader of *The Wall Street Journal*—say, a civic-minded executive—would be comforted to hear that concerns about global warming are overstated, especially given the tagline that identifies Balling as director of the Office of Climatology at Arizona State University. I doubt that the same reader would be quite as sanguine, however, if he knew that some of Balling's work was underwritten by German and British coal interests and by the government of Kuwait.

Among the skeptics, Dr. S. Fred Singer stands out for being consistently forthcoming about his funding by large oil interests. On a 1994 appearance on the television program *Nightline*, Singer did not deny having received funding from the Reverend Sun Myung Moon (to whose newspaper, *The Washington Times*, he is a regular contributor and whose organization has published three of his books). Nor has he apologized for his funding from Exxon, Shell, ARCO, Unocal, and Sun Oil. Singer's defense is that his scientific position on global atmospheric issues predates that funding and has not changed because of it.

This interesting point raises an equally interesting question. What would happen if the climate skeptics just happened to stumble on a piece of evidence confirming that global warming is indeed intensifying? Would they be willing to alter the direction of their research at the risk of cutting off their industry funding? Such a situation, to say the least, would provide them with a very serious personal and professional conflict of interest. Fortunately for Singer, Michaels, and Balling, such a situation has never apparently arisen.

In early 1995, several months before one round of international climate negotiations, Singer proposed to an oil industry public relations outlet a $95,000 project in which he would mount a series of panels, lectures, and publications to "stem the tide towards ever more onerous controls on energy use." The project was intended to publicly counter the findings of the IPCC.

Indeed, the project bears more than a faint resemblance to the coal industry's ICE campaign. Singer's proposed oil-company-sponsored public education program would prepare a "scientifically sound and persuasive critique of the IPCC summary. . . . Next we would distribute the Critique widely . . . and publish a Statement of Support by a hundred or more climate experts. This Statement could then be quoted or reprinted in newspapers. Our proposal envisages assembling a panel of about five distinguished scientists/technologists. This panel would issue a Release pointing up the IPCC Critique and conduct press briefings to defend its conclusions. If funding can be provided without delay, the panel . . . could issue its Release . . . during or before the [round of international climate negotiations] meeting in New York."

In his wind-up, Singer warned the oil companies that they face the same threat as the chemical firms that produced chlorofluorocarbons (CFCs), a class of chemicals that were found to be depleting the earth's protective ozone layer. "It took only five years to go from . . . mandating a simple freeze of production [of CFCs] at 1985 levels, to the 1992 decision of a complete production phase-out—all on the basis of quite insubstantial science," Singer wrote.

Contrary to his assertion, however, virtually all relevant researchers say the link between CFCs and ozone depletion rests on unassailable scientific evidence. Three months later, as if to underscore the CFC-ozone connection, the research director of the European Union Commission announced that the previous winter's ozone loss would result in about 80,000 additional cases of skin cancer in Western Europe. Shortly thereafter, the three scientists who discovered the CFC-ozone link were awarded the Nobel prize for chemistry. But that did not faze Singer, who proceeded to attack the Nobel committee in the pages of Reverend Moon's *Washington Times.* In his November 1995 article, headlined "Ozone Politics with a Nobel Imprimatur," Singer declared that "the Swedish Academy of Sciences has chosen to make a political statement. . . . The selection committee evidently decided to reward global environmentalism rather than a fundamental advance in the basic science of chemistry."

The following month, researchers for the World Meteorological Organization announced that the ozone hole over Antarctica had grown at an unprecedented rate in 1995, covering an area twice the size of Europe at its peak in October. A one-percent-per-day decline in the ozone layer during August had caused the ozone hole to expand more rapidly than in any previous year, reaching a maximum of 7.7 million square miles, the researchers reported.

Singer's tantrum against the Nobel committee would be laughable—except that his views exert serious influence, especially on conservative politicians. Based in part on Singer's work, House majority whip Tom DeLay and Representative John Doolittle are making an effort to withdraw U.S. participation in the Montreal Protocol—the international compact that mandates an end to production of the chemicals that destroy the ozone layer. Despite the remarkable international consensus on the Montreal Protocol, DeLay used Singer's pronouncements to attribute it merely to "a media scare."

Not long after launching his attack on the Swedish Academy, Singer revisited the climate debate in another article in *The Washington Times,* again using the vocabulary of propaganda. "Early this year," he wrote, "*The New York Times* ominously warned ' '95 Hottest Year on Record,' implying the existence of a strong global warming trend, presumably caused by the greenhouse effect of increasing atmospheric carbon dioxide. The story did not reveal that the headline was based on an earlier British Meteorological Office release that used data from only the first 11 months of 1995. December turned out to be cold, making 1995 an average year. The scary headlines, echoed by a *Newsweek* article, created only a minor stir—appearing just as the January blizzards began to hit. Nice try, fellows: bad timing!"

Unfortunately for him, Singer made two mistakes. First, severe winter weather is perfectly consistent with global warming. One effect of climate change is to produce more extreme local temperatures—leading to hotter hots, colder colds, and more severe snowstorms. Global warming has not yet eradicated the seasons of the year, even if it may be affecting their timing. Although this point is part of our basic knowledge of climate change, it appears to have eluded Singer.

Singer was also wrong about the weather record. According to the World Meteorological Organization, even when the unusually cold December temperatures are included, 1995 turns out to have been the hottest year in recorded weather history. The World Meteorological Organization also notes that there were more hurricanes over the Atlantic than in any year since 1933 and that the Antarctic ozone hole lasted longer than in any previous year.

Despite his obvious misstatements of fact, however, Singer continues to be widely quoted by the news media.

Of all the skeptics, the most infuriating to his adversaries—and the most unassailable because of his prestigious credentials—is MIT's Richard Lindzen.

Lindzen arrived at his belief that global warming is basically a nonevent based partially on his own studies of atmospheric water vapor. Water vapor, which traps heat in the earth's atmosphere, is by far the largest gaseous contributor to planetary warming. Some years ago Lindzen theorized that atmospheric convection currents would transport water vapor through certain cloud formations into the upper atmosphere. There it would be dried out—in effect, imposing an upper limit on the vapor buildup that would otherwise have fueled atmospheric warming. Fears of a runaway greenhouse reaction, he concluded, were unfounded.

But Lindzen's theory has been contradicted by satellite and balloon observations that show that lower-level warming results in increases—not, as Lindzen predicted, decreases—in water vapor concentrations at higher altitudes. In a 1995 paper published in the *Journal of Climate,* B. J. Soden and R. Fu, researchers at Princeton University, found that according to satellite readings, areas of higher tropical convection are associated with increased stratospheric water vapor and higher greenhouse trapping, which also contradicts Lindzen's theory.

Lindzen has an excruciatingly argumentative style that at times seems relentlessly obscurantist and self-contradictory. Testifying before a Senate committee in 1991, he jousted at length with then-Senator Al Gore over his theory that the drying of upper-level water vapor would produce a cooling effect and counteract atmospheric warming.

Gore asked Lindzen if he had in fact postulated a drying of the upper atmosphere. Lindzen responded, "I am saying the subsidence does produce drying, but you have a balance between the source of moisture there and the drying due to subsidence."

> GORE: "Okay, now, but wait a minute. . . . The Lindzen mechanism, if it is proven to be correct, will produce a drying of the upper troposphere. Correct? Right or wrong?"
>
> LINDZEN: "This has been hypothesized by us." . . .
>
> GORE: "Okay. Now do we have any evidence that it is, in fact, occurring?"
>
> LINDZEN: "The answer is yes."
>
> GORE: "Okay. What is it?"
>
> LINDZEN: "Okay. There is one, and all the evidence on this, on either side, is highly controversial."
>
> GORE: "Well, before you disparage other people's evidence, could you give—"
>
> LINDZEN: "No, no. But let me present this, because I am not disparaging others. I am disparaging my own. I am saying that there is a phenomenon that has been observed . . . in the reconstruction of the climate 18,000 years ago, in the midst of the last glaciation, and this is a remarkable phenomenon whereby their measurements show—"
>
> GORE: "Excuse me. Excuse me. Dr. Lindzen, just to keep this thing from flying off the edges because of its complexity, rather than looking at 18,000 years ago, do we have any evidence, observational evidence, of the climate system today wherein there is a drying in the upper troposphere?"
>
> LINDZEN: "We have no evidence on either side for today—"

GORE: "Well, hold on just a second. Are you withdrawing the original Lindzen hypothesis?"

LINDZEN: "Oh, yes."

GORE: "Oh, you are? Okay."

LINDZEN: "Yes."

GORE: "All right. Now I just want to clarify that for the record. I want to underline it. The original Lindzen hypothesis is formally, here today, at 11:45 A.M., withdrawn. All right?"

LINDZEN: "Well, in fact, it was much earlier that we withdrew it. There was a paper—"

GORE: "Excuse me for not noticing it earlier."

LINDZEN: "Okay."

Four years later, a team of researchers with the NOAA announced that direct measurements of the lower stratosphere over Boulder, Colorado had revealed "significant amounts" of water vapor, intensifying concerns about the buildup of greenhouse gases. In an article in *Nature,* the researchers noted that the vapor concentrations "are expected to be representative of the stratosphere over the . . . northern mid-latitudes."

Lindzen dismissed the NOAA findings, saying they represented conditions only over the Boulder area. They could not be extrapolated, he said, because of uneven distribution patterns of stratospheric vapor. But in a follow-up call to the NOAA, one of the authors of the report emphatically confirmed the scope of the findings. Lindzen's opinion notwithstanding, the vapor buildup is generalized over all the northern mid-latitudes, he said.

In 1995, four years after this public retraction of his vapor theory, Lindzen told the *Minneapolis Star-Tribune* that contrary to IPCC findings, greenhouse warming will not occur because of the cooling effect of water vapor.

As a scientist, Lindzen frequently frustrates listeners with his obfuscatory pronouncements and with rebuttals that sidetrack debates into progressively narrow areas of expanding complexity. It is the mirror opposite of Lindzen the social analyst.

In August 1995 Lindzen invited me to his home, where we talked for two hours. Both he and his wife are exceedingly gracious and hospitable people. In contrast to his often tortured scientific pronouncements, I found his social and political expressions to be lucid, succinct, and unambiguous. Indeed, I found him to be one of the most ideologically extreme individuals I have ever interviewed.

His background includes service on the advisory board of the George C. Marshall Institute. Founded in the 1980s, as we have seen, the institute conducts no original research. It initially issued reports promoting President Reagan's "star wars" defense program, and more recently it has issued several reports dismissing climate change. Most researchers view these reports as ideological rather than scientific contributions. Their suspicions are understandable, given that the Marshall Institute receives support from five extremely conservative political foundations.

Lindzen shares with the Marshall Institute an ongoing crusade against the consensus findings of the Intergovernmental Panel on Climate Change. Like many ideologues, he has resorted to unwarranted ad hominem attacks on his scientific adversaries. In a striking piece of testimony in St. Paul, he asserted that the co-chair of the IPCC's scientific group, Sir John Houghton, revealed in a recent book that "he is motivated by a religious need to oppose materialism." The examining attorney produced a copy of Houghton's book and offered it to Lindzen. Rather than point out the passage himself, Lindzen referred his questioners to two sections of the text. The next day, after Lindzen had departed St. Paul, attorneys introduced a copy of those sections into evidence. Neither they nor anything else in the book came close to supporting Lindzen's accusation.

During his St. Paul testimony, similarly, Lindzen stated that Bert Bolin, the Swedish physicist who headed the IPCC, confessed to having altered summaries of the group's scientific deliberations under pressure from such advocacy groups as Greenpeace. As substantiation, Lindzen cited a 1993 article in the magazine *Physics World*. When a reading of the article revealed no such confession, Lindzen changed his testimony, saying instead that he had heard the statement on a tape of a radio program that he had forgotten to bring to St. Paul.

Lindzen and other naysayers repeated charges that the IPCC distorts the findings of its scientists and suppresses their dissents. The charges resurface in various contexts at about nine-month intervals. In 1994 the director of the Marshall Institute, Frederick Seitz, accused the authors of IPCC scientific summaries of "exaggerating risk . . . solely—we suspect—to satisfy an ideological objective of aggressively constraining the use of energy." Seitz cited three surveys of IPCC scientists, including one by S. Fred Singer, alleging "there is no consensus among climate experts in backing the assertions of the IPCC. . . . It is therefore presumptuous and quite out of order for the IPCC to claim such a consensus when all of the evidence points the other way."

To which Bert Bolin, chair of the IPCC, responded: "It seems to us the information which has been provided you regarding the IPCC is selective and in many regards inaccurate. . . . All the summaries have been agreed at these plenary meetings without dissent and none of us has received any subsequent letters of complaint from scientists regarding the final version. This thorough and completely open process has guaranteed that the summaries have had wide ownership; in no way can they be described as being the work of a select few. The process provides justification for the description of substantial scientific consensus."

Describing the process as a consensus, Bolin continued, "does not mean that all scientists are agreed on all aspects, but it does mean there has been broad agreements regarding the statements we have made. . . . The surveys of IPCC scientists to which you refer have all been very incomplete and are not a good way to measure consensus. However, it is interesting that, when asked to make their best estimate of future climate trends, the respondents tended to agree with the statements in the IPCC reports."

In conversation, Lindzen takes pains to distance himself from Michaels and Balling. He told me that Michaels comes to the climate debate from the "scientific backwater of climatology. He doesn't really know physics and he should." He attributed what he called Balling's "crude" understanding of climate dynamics to his training as a geographer, and he criticized his book as actually supporting the very computer models it purports to discredit.

Lindzen has been a paid consultant for major oil and coal interests. His 1991 trip to Washington to testify before Gore's committee was paid by Western Fuels. He addressed a meeting of OPEC delegates in Vienna in 1992 and an industry lobby group in New Zealand in 1995. Lindzen told me that he charges $2,500 a day to consult for fossil fuel interests, but for his St. Paul testimony he charged Western Fuels less than that rate, since it required ten days of preparation. Overall, Lindzen estimated, he makes around $10,000 a year on consulting.

Lindzen's ideological extremism emerges during his forays into social analysis. He asserted in an interview that the environmental movement conforms to the same sociological criteria—in terms of group behavior and dynamics—as the Nazi movement in Weimar Germany, absent the anti-Semitism. He compares the environmental movement to the eugenics movement of the early 1920s, saying both movements were built on flawed science that, in turn, gave rise to destructive policies. (Several months after this interview, the parallels between environmentalists and eugenicists formed the theme of a column by William F. O'Keefe, chairman of the Global Climate Coalition.)

In a remarkable statement, Lindzen declared that even if a relatively dire global warming scenario—with its profusion of positive feedbacks—occurred, "the negative impact on civilization would be far less significant than such major social and economic changes as the Russian Revolution."

In his 1992 address to OPEC, Lindzen again switched roles from meteorologist to social scientist. He warned his audience that predictions of global catastrophe contribute to "a societal instability . . . which can cascade into massive economic and social consequences." He attributed that instability to "the existence of large cadres of professional planners looking for work, the existence of advocacy groups looking for profitable causes, the existence of agendas in search of salable rationales, the ability of many industries to actually profit from regulation."

By contrast, IPPC chairman Bolin is a conservative scientist whose public speech is meticulously cautious and invariably understated. It is all the more conservative, given the intersection of economic and political pressures that converge on the IPCC. Despite his reluctance to engage them, Bolin recently found himself forced to respond to the conflict generated by the industry-financed skeptics. "The public press," he said, "is anxious to seize on scientific controversies, and rightly so. But an accurate account is seldom given of how well various differing views are founded in the scientific-expert community. An increasing polarization of the public debate that has been developing in some countries is not a reflection of a similar change among *experts at work* on these issues."

Reporting to international negotiators in Geneva in the summer of 1996, Bolin again addressed the skeptics and their ideological allies—this time, he took up accusations they raised in a recent Marshall Institute report, and in a paper on global warming. Bolin emphasized that their objections to the IPCC's findings "have not been subject to the careful scientific/technical review that is a standard procedure for publication in regular . . . journals, and also for IPCC reports." Nevertheless, he added, "many such comments have nonetheless been considered and most of them rejected in the IPCC process because of inadequate scientific bases."

Simply because a person dissents does not mean he is wrong. The respectful consideration of dissenting views is critical to an open, democratic, scientific process. But in the case of the greenhouse skeptics, the near unanimous professional judgment of their IPCC peers has rejected their views.

That would be the end of the skeptics' relevance to the climate debate—except that big oil and big coal use them to the hilt.

How much money the fossil fuel industry spends on disinformation is impossible to determine, given the right of secrecy that corporations enjoy. Only their nonprofit subsidiaries afford the public a glimpse of their financial activities. In response to a series of Freedom of Information Act requests, the Internal Revenue Service provided documents that indicate that in 1994 and 1995 the Global Climate Coalition spent more than a million dollars to downplay the threat of climate change. The GCC projected it would spend nearly another million on the issue in 1996. Each year, U.S. businesses spend some $500 million to "greenwash" their corporate images, according to the nonprofit journal *PR Watch*. It is not known how much of that aggregate public relations budget is devoted to downplaying climate change. But a clue can be found in the tax filings of the National Coal Association. In 1992 and 1993 the association spent more than $700,000 on global climate efforts. Similarly, the American Automobile Manufacturers' Association in 1993 spent nearly $100,000 on "global climate change representation" and on membership dues to the GCC.

In 1993 alone, the American Petroleum Institute (API), just one of fifty-four industry members of the GCC, paid $1.8 million to the public relations firm of Burson-Marsteller—a firm credited by former Treasury Secretary Lloyd Bentsen with spearheading the defeat of a proposed tax on fossil fuels. By comparison, the five major environmental groups that focus on climate issues have a combined annual expenditure only slightly greater than the API's one-year public relations allocation of $1.8 million. According to figures provided by officials of the Environmental Defense Fund, the Natural Resources Defense Council, the Sierra Club, the Union of Concerned Scientists, and the World Wildlife Fund, their annual spending added together totals about $2.1 million.

Several of the skeptics—especially Michaels and Singer—have expressed a belief that the federal government should get out of the research business altogether. They call for a privatized research establishment, funded by those industrial interests that are most directly affected by the research. The obvious potential for conflicts of interest seems to have escaped them.

As the country's first administrator of the Environmental Protection Agency, William Ruckelshaus has firsthand knowledge of both the benefits and the limitations of government's role in funding research. Presently chief executive officer of one of the country's largest waste disposal firms, Browning-Ferris Industries, Ruckelshaus is also quite conversant with the tendencies of industry. "If the critical questions about climate change," he said in an interview, "involve gaining a better understanding of what impacts to expect and what rate of change we might see, then the government must be the entity funding the research. At this point, we urgently need much more research on the nature of the problem. Unfortunately, to the extent that industry funds scientists, it does so to debunk the IPCC. Anyone who happens to be a shareholder in one of those companies should tell that company to go spend their money on something else.

"Once scientists have described the problem and policies are proposed to deal with it, we know that if those policies will adversely affect a specific industry, they're most likely to resist. It's in the nature of the system to happen that way. They'll go through denial. They will continue to refute evidence that their activities are causing these problems. It is clear the government has to take the lead by providing the research."

Government-funded science provides an assurance of integrity. Conducted according to the exacting standards of professional academic research, it is the arena where truth has the greatest chance, however small, of being revealed. To compromise the standards of scientific research would be to open the door to a direct assault on society's most highly developed body of knowledge. Under the pressure of advanced techniques of marketing, promotion, and sales, our standards in other areas have already buckled.

If the fossil fuel industry were honestly concerned about getting at the truth of climate change, it would contribute to a blind-trust pool of private research funds that the federal government would allocate to researchers, with guidance from bodies like the National Science Foundation, the National Academy of Science, and the National Research Council. Strangely enough, despite the vociferous defense of private funding by oil and coal lobbyists, none has volunteered to fund such a pool.

Given the accumulating evidence that the burning of oil and coal is threatening the planet's climate, why do the dismissals by Lindzen, Michaels, Balling, Singer, Idso, and the other skeptics carry such weight with reporters?

One answer lies in the ethical standards of journalism. The

professional canon of journalistic fairness requires reporters who write about a controversy to present competing points of view. When the issue is of a political or social nature, fairness—presenting the most compelling arguments of both sides with equal weight—is a fundamental check on biased reporting. But this canon causes problems when it is applied to issues of science. It seems to demand that journalists present competing points of view on a scientific question as though they had equal scientific weight, when actually they do not. The problem escalates because most journalists are not qualified to make judgments about issues such as standing, expertise, and integrity within the scientific community. As a result, ideology disguised as science can contaminate the debate.

For reporters who are unable to educate themselves in the scientific field that they cover, there is a test of sorts. It is partial and inadequate but worth considering: *A scientist's authority ends at the boundary of his professional expertise.* In areas of policy and values, the scientist has no more authority than the rest of us. All our opinions are of equal weight, insofar as they conform to what we know. Precisely because scientific expertise is limited to issues of science, we must all be allowed to debate policies and values. And we must be permitted to do it openly, honestly, without the secret manipulation of a one-interest industry, in the well-lit public arena.

Dr. Stephen Schneider of Stanford University, a pioneer and a prominent commentator on global warming, wrote several years ago: "the proper role of a scientist in a policy debate is not to claim any special ability to be able to choose the 'right solution.' The professional role of the scientist . . . is to state what can happen and what the probabilities are. The professional scientific role ends there, and anything beyond that dips into a personal values realm."

At a hearing last November, one of the most distinguished American climate researchers, Jerry Mahlman, told an assembled congressional audience: "Because I speak with credentials as a physical scientist, I do not offer personal opinions on what society should do about these predicted climate changes. Societal actions in response to greenhouse warming involve value judgments that are beyond the realm of climate science. Indeed, I would encourage your skepticism whenever you hear a climate scientist's prediction that is accompanied by a policy opinion." This attitude—not the diatribes of Michaels or the ideological pronouncements of Lindzen—reflects the scientific conscience.

But that same restraint—the one that requires an internal wall between one's scientific obligations and one's personal beliefs—puts genuine scientists at a serious disadvantage in the arena of public debate. "Scientists, by nature, are very conservative," explains James McCarthy. "They don't want to overstate things. Rather than express themselves in terms of absolutes, they couch their statements in terms of statistical probabilities. In this debate, the skeptics are atypical of the general scientific community. The irony is that when you find scientists who are adamant on any side, they look very curious within the scientific community. But to the general public, they come across as the most knowledgeable and authoritative of all."

A reporter from *Der Spiegel,* the German newsmagazine, put it more succinctly. After an article disclosing the industry funding of several skeptic scientists appeared in *Harper's* magazine, the *Spiegel* reporter called me in late 1995 to ask why the German Coal Mining Association would be funding the research of Pat Michaels and Robert Balling.

"Seems pretty obvious to me," I replied. "They have a direct interest in the outcome of the debate."

The reporter responded, "It's only in America that there is a debate over what's happening to the global climate. In all the European scientific circles, there's no debate at all about what's happening. Except for your country, the only debates are how fast and with what impacts the changes will happen."

Unfortunately, his self-assurance was short-lived. In March 1996 simultaneous press conferences were held in Geneva and London announcing the formation of the European Science and Environment Forum (ESEF)—the European branch of the skeptics.

The ESEF's positions echo those of Western Fuels, the Marshall Institute, and the greenhouse skeptics:

- "The IPCC undermines its scientific integrity by condoning over-simplified summaries of extremely complex studies that can be easily misinterpreted."
- "The IPCC's predictions of global climate change are generated from models which are based on crude approximations, have an inadequate physical basis, and exclude several extremely important factors."
- "Any benefit of climate change is usually ignored in favor of pessimistic visions."

At the press conference in Geneva, three of the group's founders were asked whether their work was funded by industry. One, Roger Bate of Cambridge University, said that he got a mix of public and private funding. The other two members declined to answer the question.

A list of founding members of the ESEF revealed the names Patrick Michaels, Robert Balling, and S. Fred Singer.

By the spring of 1996, in parts of North Korea, people were reduced to eating leaves, grasses, and wild roots following the most extreme floods in memory. In April North Korean officials renewed their plea for emergency international food aid. Spokesman Pak Dok Hun said the country urgently needed 1.2 million tons of grain by the following October. The devastating floods, according to officials, destroyed not only the 1995 rice harvest but also rice stocks from previous years, as well as homes and factories, and it flooded mines in western portions of the country. Over 5 million people, out of a population of 23 million, were said to be affected, most by food shortages.

Halfway across the world, the Blue Hills meteorological station outside Boston recorded 11 inches of snow on April 10, 1996. That storm brought the total snowfall in Boston for the winter of 1995–96 to 140 inches, the highest in the city's recorded history. The previous winter had seen near record-low snowfalls, which, in turn, had followed a near-record snowfall of about 110 inches in 1993–94.

At the end of April 1996, in parts of southern Indiana and Illinois, authorities braced for what they expected to be the worst flooding in 30 years as a foot of fresh snow fell on northern Michigan and six inches of April snow covered much of Wisconsin. The latest snowfall set a season record at Marquette, Michigan. On Grand Traverse Bay in northern Michigan, the water froze over at the beginning of February and did not melt until April 20—the longest freeze in 70 years, according to Coast Guard officials.

THREE

A Congressional Book Burning

SCIENCE AND POLITICS DON'T MIX. SCIENTIFIC PROOF evolves through exacting methodological standards, the replicability of experiments, and the intense scrutiny of other scientists in the field—the process called peer review. Political battles, by contrast, are determined by the rule of the numerical majority, which, as history shows us over and again, is not always the test of wisdom.

Inherently, science and politics are incompatible, but science is essential to the formation of government policy in many areas. So Congress many years ago established the Office of Technology Assessment (OTA)—a nonpartisan body of specialists to screen and filter the science on which the policymakers depend. The OTA surveyed relevant scientific fields to determine the weight of accepted evidence. It also evaluated the professional credentials of experts to determine which were worthy of congressional attention.

In an ominous but little-publicized move following the election of a large number of highly conservative and inexperienced

Republican congressmen in 1994, Congress abolished the OTA. With that stroke of political expression began the politicization of science and, in the area of climate change, the domination of a most willful and determined ignorance.

Congress's assault on the scientific evidence for planetary warming was preceded by a House subcommittee's attack on the scientific evidence for ozone depletion—a body of research in which the areas of scientific uncertainty are far smaller than they are in climate science.

The earth's protective layer of atmospheric ozone shields life from the destructive ultra-violet B radiation from the sun. But substantial amounts of ozone have disappeared from the atmosphere over Antarctica, as well as over Greenland, Scandinavia, and western Siberia. That the ongoing depletion of the ozone layer threatens both human life and biological systems has been accepted by chemical manufacturers and governments alike. Since 1987, more than 150 governments have signed the Montreal Protocol—an international agreement to phase out the production and use of the chemicals that are depleting the ozone layer.

But in September 1995, despite the depletion, the House Science Subcommittee on Energy and the Environment began a series of hearings on what it called "Scientific Integrity and the Public Trust," the first of which revisited the already-proven issue of ozone depletion.

From his opening statement, subcommittee chair Dana Rohrabacher left no doubt as to where he stood on the issue. The ozone scare, he said, has "turned out to be another basically the sky-is-falling cry from an environmental Chicken Little, a cry we've heard before when the American people were scared into the immediate removal of asbestos from their schools. . . . This time the scaremongers managed to stampede the Congress and President [George Bush]." In an Orwellian inversion of reality, Rohrabacher continued, "The American people deserve better of their government than scare tactics that are designed to intimidate and repress rational discussion."

Picking up where Rohrabacher left off, Representative John Doolittle, another California Republican, questioned whether the research findings justified a ban on ozone-depleting chemicals. "My own belief," Doolittle proclaimed, "is that the question is still very much open to debate. . . . Theories or speculation about this are not sufficient. We need science, not pseudo-science."

Doolittle was immediately challenged by a Democratic member of the subcommittee, Representative Lynn Rivers of Michigan, who asked Doolittle whether he is a scientist. When he replied in the negative, she asked: "Have you found in peer-reviewed articles or in the broader scientific discourse that people are saying this is not really a problem?"

> DOOLITTLE: "I have found that there is no established consensus as what actually is the problem. I found extremely misleading representations by the government and government officials that are not founded on sound science."
>
> RIVERS: ". . . [W]hat I was asking about is peer-reviewed articles [by] scientists who are . . . doing this work on a regular basis. Can you give me an example of some of the peer-reviewed publications that you consulted in formulating your opinion that there's no [sound] science?"
>
> DOOLITTLE: "Well, you're going to hear from one of the scientists today, Dr. Fred Singer."
>
> RIVERS: "Dr. Singer doesn't publish in peer-reviewed documents."
>
> DOOLITTLE: "[I]'m not going to get involved in a mumbo-jumbo of peer-reviewed documents. There's politics within the scientific community, where they're all intimidated to speak out once someone has staked out a position. . . . And, under *this* Congress, we're going to get to the truth and not just the academic politics."
>
> RIVERS: "[T]he general way to feel certain that you're getting good science is that you put your ideas out in a

straightforward way in a peer-reviewed publication and you allow others who are doing the same work to make comments, to criticize, to replicate your findings. And what I'm asking you, in your search for good science, is what peer-reviewed documentation did you use to come up with your decision? What good science did you rely on?"

DOOLITTLE: "My response to you is, it is the proponents of the ban [on ozone-depleting chemicals] that have the burden of producing the good science. I do not have that burden."

Shortly thereafter, House majority whip Tom DeLay changed the subject by asserting that switching to ozone-protecting chemicals would be less energy efficient and therefore not cost-effective. Taking his cue from DeLay, Rohrabacher displayed his qualifications to head a scientific subcommittee of Congress: "When you say the energy requirements on the alternatives are increasing, and would increase, the necessity to use more fuel, what you are actually saying then is more carbo . . . they're not carbohydrates. Carbohydrates is what you eat."

DELAY: "Hydrocarbons."

ROHRABACHER: "Hydrocarbons are going into the atmosphere."

In 1994 the scientific community had published an extensive assessment of ozone depletion. The report represented a consensus of thinking shared by virtually all the relevant researchers in the field. Asked about his reaction to the assessment, DeLay replied that he had not consulted it before formulating his position. Asked why not, DeLay responded: "Well, I just haven't been presented with the study of late. I'll be glad to read it and give you my assessment of it."

DeLay instead cited *Toxic Terror* by Dr. Elizabeth Whelan, of the American Council on Science and Health. This book, in fact, excoriates environmentalists for exaggerating environmental health risks putatively to scare the American public. But as *Washington Post* reporter Howard Kurtz noted in an article in the *Columbia Journalism Review,* Whelan, who praises the nutritional value of fast food, receives money from Burger King. While receiving funding from Oscar Mayer, Frito Lay, and Land O' Lakes, she dismisses links between fatty diets and heart disease. According to John Stauber, editor of *PR Watch* and author of a recent book exposing extensive environmental misrepresentations of the public relations industry, she attacked a crusade against the use of fatty palm oils in movie-theater popcorn while she was in the pay of palm oil producers.

Led by Rohrabacher and DeLay, the Republican majority on the subcommittee introduced their big guns in the case against ozone science—Fred Singer and several witnesses from ideologically conservative think tanks, including the Competitive Enterprise Institute and the George C. Marshall Institute. In elevating political opinion over scientific fact, the congressmen ignored findings from balloon and satellite studies that have documented significant decreases in atmospheric ozone and shown that the Antarctic ozone holes in 1990 and 1992–94 were the most severe on record.

Probably because of the unassailable certainty of ozone science—as well as its endorsement by both the Bush administration and the chemical industry—the subcommittee eventually pulled in its horns on that issue. Some speculate that the order to retreat came from House Speaker Newt Gingrich who, realizing that the assault on ozone science was a lost cause, urged the subcommittee to focus instead on climate change.

The climate hearings revealed the Republican leadership's eagerness to attack mainstream climate scientists and to subordinate their findings to those of the industry-sponsored scientific skeptics. "I am not swayed," Rohrabacher said, "by arguments that here's a big list of scientists that are on my side and you have a smaller group of scientists on your side. I'm just not swayed by that at all." This from a congressman whose scientific training barely allows him to distinguish between carbohydrates and hydrocarbons.

Rohrabacher seems to share with many of his young, ideological colleagues what may be called a Galileo complex—an emotional identification with a martyred and shunned voice-in-the-wilderness who eventually attains redemption for his heroic, lonely battle. It is a mindset that the true believers in the House share with extremists elsewhere.

When Rohrabacher conducted the "Scientific Integrity and the Public Trust" hearings on climate change in the fall of 1995, he had no qualms about substituting his own ideological standard for professional scientific standards. At those hearings he gave Pat Michaels—recipient of thousands of dollars from the coal and oil interests and from the leading corporate funder of the Wise Use movement, editor of a non-peer-reviewed journal, and an initial participant in the cynical ICE campaign—equal billing with one of the most accomplished climate scientists in the country—Jerry Mahlman, director of the NOAA's Geophysical Fluid Dynamics Laboratory and chair of NASA's Mission to Planet Earth Scientific Advisory Committee.

Rohrabacher's announced theme—"Scientific Integrity and the Public Trust"—translated into automatic distrust of all government-funded scientists who, the congressman is fond of saying, are motivated to create or exaggerate problems in order to secure government funding. In particular, the hearings focused on whether the House Science Committee should continue to fund the U.S. Global Change Research Program, the largest single government research program on climate change. The program had a $1.2 billion budget that was distributed among researchers at NASA, NOAA, and other agencies. The committee also contemplated the funding of climate change programs at the Department of Energy and the EPA. According to Rohrabacher, the federal research effort on climate change was "a product basically of the Vice President of the United States' zeal for this particular issue. Many of us believe that zeal is what we could call environmental fanaticism." Opening the first panel of the hearing—which focused on the state of climate computer models—Rohrabacher continued to politicize the scientific process by asserting that the issue of climate change was an expression of Vice President Gore's liberal political agenda.

Unintimidated by Rohrabacher, Jerry Mahlman addressed the scientific probabilities and uncertainties with equal care. He issued an implicit moral challenge to the skeptic scientists—Singer, Michaels, Balling, and Lindzen. He warned the Congress that it should separate science from policy: "societal actions in response to greenhouse warming involve value judgments that are beyond the realm of climate science. Indeed, I would encourage your skepticism whenever you hear a climate scientist's prediction that is accompanied by a policy option." He noted that the IPCC assessments are "the most widely accepted statements ever on climate change" and recommended the legislators use them as "a point of departure for evaluating the credibility of opinions that disagree with them."

Large areas of uncertainty remain concerning the impacts and rates of climate change, Mahlman conceded. But he noted that "none of the uncertainties I will discuss here can make current concerns about greenhouse warming go away." Speaking in the accepted scientific language of probability, Mahlman gave his assessment of the various effects: "Future surface warming is very probable." "Rise in sea level very probable." "Summer midcontinental dryness and warming probable." "Increased tropic storm intensities uncertain. A warmer, wetter atmosphere could indeed lead to increased intensities of tropical storms, such as hurricanes. Presently this problem has not been properly addressed due to inadequate computer power and to uncertainties in regional climate change predictions."

Concluded Mahlman: "In summary, the greenhouse warming is quite real. The state of the science is strong . . . it is a virtually certain bet that this problem will refuse to go away no matter what is said or done about it over the next five years."

Rohrabacher countered Mahlman by bringing on his star witness, Pat Michaels, who wasted no time in asserting his own beliefs. In terms that would play to the apparent emotional propensities of Rohrabacher and many other Republican congressmen, Michaels identified himself as a heroic dissenter unjustly victimized by a conspiracy of climate scientists. "For the last decade, a . . . small minority [of scientists] has argued that, based upon the data on climate change, the modeled warming was too large, and therefore any intrusive policy would not be based on reliable models of global warming," he testified. "This view has been cast in a very negative political light which has had a chilling effect on scientific free speech."

Attacking the IPCC assessments, Michaels declared that "the

scientific review process [on which the UN climate negotiations are based] has been highly flawed . . . the good news . . . is that the so-called skeptics were right." With that self-congratulatory flourish, he blithely dismissed what is probably the most inclusive review process in history—one that actively invites the scrutiny of every relevant practicing scientist in the world.

Rohrabacher went on to solicit Michaels's view of government research funding. The scientist replied in his role as a fiscally conservative bureaucracy basher. "I have not seen the fine print," he responded. "In Washington, there is a lot of fine print that spends a lot of money. So the only answer I can give to you is, if I saw the fine print, I would give you an opinion."

Warming to Michaels's attack on bureaucracy, Rohrabacher said, "I am taking a look here [at an] $111 million contract for global climate research [by the] EPA, and their account was used for brochures, posters, program logos, design for product awards, promotional pens, pencils, buttons, banners, displays, billboards, bus and train placards. That is the fine print that you are talking about, Dr. Michaels."

That leaders of the world's largest insurance companies are issuing a stream of alarms about the potential bankruptcy of their industry didn't faze Michaels. Later in the hearings, he propelled his policy pronouncements into full flight, asserting: "There has been no significant change—in fact, it has been a significant decline in the intensity of Atlantic hurricanes over the last fifty years, regardless of what the insurance agency says. I know they like to raise rates." But if Michaels omits to read the fine print, he seems positively oblivious to headlines. Industry data show that since 1987, fifteen severe storms have cost $50 billion in insurance claims. Hurricane Andrew alone involved over $16 billion in insured losses. In fact, while weather-related disasters in the 1980s cost the insurance industry just over $50 billion, they have cost the industry $162 billion for the first five years of the 1990s alone. Michaels did not mention that 1995 was the most active hurricane season in the Atlantic since 1933—or that the increase in tornadoes is associated with the same increase in severe storms.

Dismissing the consequences of climate change, Michaels concluded by assuming the mantle of ecologist, noting: "We are . . . approaching a new paradigm of a new view of the world, which is going to change from fragile earth . . . [to] the concept that the earth is more resilient than we had once feared it might be."

Mahlman, when he testified again, persisted in his attempt to separate science from rhetoric: "I said . . . I did not think it appropriate for climate scientists to offer political or sociological opinions. . . . But I do feel it is appropriate for climate scientists, such as myself, to speak to what the problem is. The problem is that global warming is something that is a harsh and inexorable reality. We do not know, quite for sure, whether it is at the lower end or the upper end of the range. . . . Pat Michaels's arguments do not resolve that debate. . . . But we do know that it takes a long time to build up carbon dioxide levels that are high enough to be scary. We also know that . . . it takes a very, very long time for them to go away."

Distinguishing his personal policy opinion from his scientific testimony, Mahlman continued: "It seems reasonable to me—as a public citizen—that the issue has many aspects that are . . . fraught with complexities, not just on the climate side but on the impacts side. . . . The problem is difficult, uncertainties are significant, the cost of doing something about it, particularly if you grew up in a coal-producing part of the country, is very large. . . . The other side of the same coin, however, is that the cost of not doing something about it may be prodigious. . . . [T]he problem may last for hundreds to maybe even a thousand years. . . . There is no soft landing spot independent of one's rhetoric or one's political position." In the end, Mahlman's testimony was wasted on Rohrabacher.

During the second panel of the hearings—which focused on impacts of climate change—the subcommittee heard testimony from David Gardiner, an assistant administrator of the EPA. Discussing the EPA's estimate of sea level rise, Gardiner noted that such a rise "could drown [up to] 60 percent of our coastal wetlands." It would, he said, "inundate more than five thousand square miles, an area the size of Connecticut, of dry land in the United States if no protective actions are taken." He added that "increasing sea level poses significant risks to our environment. Some of those effects are potentially catastrophic and irreversible." It was at this point that Rohrabacher, an avid California surfer, responded: "I am tempted to ask what this will do to the shape of the waves and rideability of the surf. But I will not do that. I wait until later when we get off the record."

Later in the hearing, Dr. Robert Watson, the lead author of the IPCC's 1995 report on impacts of climate change—and, at the time, a senior scientist in the White House Office of Science and Technology Policy—testified. (In September 1996 Watson was elected by a unanimous vote of his colleagues to replace Bert Bolin as chair of the IPCC.) Rohrabacher questioned Watson about the finding that one-third of the world's forests will be lost or altered as a result of climate change.

> WATSON: "That particular conclusion comes from a complex forest ecological model, where it takes plausible changes in temperature, precipitation, evapotranspiration, and soil moisture, and looks to see to what degree . . . species move, how its ecosystems disassemble and reassemble. . . ."
>
> ROHRABACHER: "And this is also true of the projection that rangelands will be lost?"
>
> WATSON: "Yes, based on ecological modeling."
>
> ROHRABACHER: "And that desertification will increase?"
>
> WATSON: "You don't even need a model to come to that conclusion. The answer is quite clearly yes."
>
> ROHRABACHER: "Do models project that one-third to one-half of existing mountain glaciers will disappear?"
>
> WATSON: "That comes from simply understanding the relationship between mountain glaciers and their thermal structure.". . .
>
> ROHRABACHER: "Let me ask you, Dr. Watson, you suggested that there was a one-half-degree-centigrade temperature rise in this century? Is that what you testified?"

WATSON: "That . . . has been a consistent view of the IPCC."

ROHRABACHER: "Now, I will have to say that I found . . . Patrick Michaels's, Dr. Michaels's [testimony] to be absolutely understandable and devastating to your argument."

WATSON: "What we have is hundreds of thousands, if not millions, of data points on the surface of the earth, both land and the ocean, since 1860. And it is that database that strongly suggests . . . there is directive evidence . . . that the earth's surface has warmed 0.3 to 0.6 degrees Centigrade."

ROHRABACHER: "Yes . . . I am going to just be frank with you. I find it impossible to understand that scientists are sitting here telling me that this is as significant as you suggest." . . .

WATSON: "One needs to also look at rates of temperature change, and if you look at the paleo record, it shows the types of projections we are making . . . over the next hundred years is a rate that is unprecedented in the last ten thousand years. That is a rate that these complex forestry systems can't keep pace with."

ROHRABACHER: "Dr. Watson . . . what happens when we get these type of what I consider to be unjustified scare scenarios is that we start ignoring the economics. . . . The reason why you find skepticism from this chairman is that I have met so many people who feel absolutely justified . . . in stating a problem and then exaggerating it to the point because we have got to get public attention on this because that is their priority."

WATSON: "You don't see [exaggeration] in the international scientific assessments. The international assessments don't use the word *apocalypse*. You don't see it

> in the testimony of any of the people in front of you here."
>
> ROHRABACHER: "[A]fter hearing some of your either predictions or projections, I would have probably used this headline [of apocalypse], even though you didn't use the word *apocalypse.*"

In March 1996 the full Science Committee held a major hearing to determine its final budget allocations. The hearing provided a case study in the use of a tiny band of industry-sponsored skeptics to justify a congressional assault on mainstream science. Coincidentally, the committee called the same three skeptics who had been paid by the coal industry the previous year to testify in Minnesota that fossil fuel burning is not disturbing the global climate. To a man, Michaels, Balling, and Lindzen all agreed on the generality that more data were needed. But none supported the specific monitoring and assessment programs of the U.S. Global Change Research Program, NASA, the NOAA, the Department of Energy, or the EPA.

When his turn came, Robert Balling's testimony before the Science Committee turned into an exercise in selective omission. Balling noted that "models throughout the world tell us that the bulk of the warming should be in the northern hemisphere, in the Arctic. . . . [But] if we go back and look at the thermometer data that would be available for the Arctic for the last fifty years, we . . . see that there's absolutely no warming in the Arctic. This is quite troubling when this is the part of the planet where we think we should see the greatest amount of warming at the present time."

Balling neglected to tell legislators that the Arctic is, in fact, warming. Researchers have discovered soil temperature increases of 2 to 5 degrees Celsius in Alaska during this century, as well as the warming of a deep layer of the Arctic Ocean. Even more striking, a team of NOAA researchers recently found an increase in surface temperature of 5.5 degrees Celsius (about 9 degrees Fahrenheit) over the last thirty years at nine stations north of the Arctic Circle. In other words, the warming is occurring precisely where the models say it should be—and the evidence directly contradicts Balling's testimony.

Next, the MIT contrarian, Richard Lindzen, took aim at the central program of Mission to Planet Earth—the EOS (Earth Observing Satellite) program. "EOS started in an unfocused way, to just take a lot of measurements and expect this would answer and, in fact, ask the questions," Lindzen told the panel. "If you simply expect that a blockbuster approach to taking data will automatically tell you answers, that's pointless. . . . EOS was starting with the instruments and hoping the questions would arise. . . . There's also a need, I think, to have some ideas on the books to which an observation provides a specific test. And there's rather little of that in EOS" in Lindzen's judgment.

Lindzen's attack on the EOS program carried serious weight with the Republican leadership of the committee. The rebuttals did not. Watson, for instance, responded, "Over the last decade, hundreds of scientists have been involved in posing questions and relating those questions to the types of measurements that can be made from EOS. . . . [I]t is a science-driven program." He cited the work of government researchers in studying how data from commercial and military satellites as well as EOS, combined with "modeling and ground-based observations, will all play together."

Predictably, the IPCC—the favored whipping boy of both Lindzen and the Marshall Institute—is also a favored target of conservative Republicans. Even apart from the climate issue, these Republicans seem terrified by anything that remotely suggests world government and are therefore deeply hostile to everything associated with the United Nations. Lindzen, for example, told the committee that the process of the UN-created IPCC is "flawed" and that, despite the very wide range of views of its contributing scientists, the IPCC "policy-makers' summary and the executive summaries tend to misrepresent this."

Again Watson was left to defend the IPPC's unusually inclusive and open scientific process. Any document used by the IPCC, he explained, either comes from the peer-reviewed literature or is circulated in preprint form for peer review. "There seems to be a misunderstanding of who writes the [IPCC] summary," he said. ". . . Most scientists agree that . . . we do indeed talk about what is known and not known and we actually do quite well describe uncertainties. . . . These are indeed summaries written by the scientists and then peer-reviewed by both the experts and by the governments themselves. So . . . the summary for policy-makers should be viewed as being written by the peer-review community, just like the main text."

Ultimately, Science Committee chairman Robert Walker of Pennsylvania praised the skeptics, underscoring their political effectiveness. "I think [they have provided] a valuable contribution," he said, "to the understanding of this Committee. . . . I happen to think I'm learning something." An early, enthusiastic proponent of the discredited findings about cold fusion, Walker would subsequently introduce legislation to substantially slash funding for research into climate change.

The real aim of the hearings became clear in May, when the Science Committee finally determined what programs to recommend for funding. Rohrabacher's subcommittee recommended that it cut NASA's Mission to Planet Earth—with its EOS program—by $400 million, or about one-third of its budget. It substantially reduced the arm of NOAA that researches warming-related changes in ocean dynamics and marine life. Finally, the committee considered the EPA's global monitoring program, which assesses the vulnerability of various ecosystems to climate change. After brief deliberation, the committee voted to defund it entirely.

Many of these cuts had been recommended by Representative Rohrabacher. They are consistent with his votes, the previous year, to allow the export of oil from Alaska's North Slope and to deny a $45 million appropriation to promote the export of American solar energy technology. But it was Walker, as chair of the Science Committee, who made the final decisions. And he made them in accordance with the counsel he received not from a scientific source but from the George C. Marshall Institute. (Eventually, the full Senate restored some funding to Mission to Planet Earth and to NOAA. That restoration notwithstanding, NOAA's program to monitor climate change lost one-third of its funding.)

In the final budget meeting of the Science Committee on May 1, 1996, Walker asked Rohrabacher for his opinion of the various government programs to monitor climate change. Rohrabacher's response reflects the contempt in which he holds this hard-won scientific evidence. "I think that money that goes into this global warming research is really money right down a rathole," he said.

"What happened between now and last year when we had . . . a smaller amendment to add more money to global warming research is that we had a very cold winter. And I remember scraping the ice off my windshield, and my nose was freezing and my ears felt like they were going to come off. And I looked up and I just said, thank God for global warming or I'd be even more miserable than I am now.

"The fact is that global warming, the more I have studied the issue, the more I have come to believe . . . that at best it's nonproven and at worst it's liberal claptrap. And in fact I have come to the conclusion more every day that it's more toward the latter than the former. We have very scarce dollars, and . . . this idea about global warming is more like a religion than it is a science.

"And you've got people who are absolutely convinced that in some way mankind is causing . . . a one- or two- or three-degree change in our temperature over a hundred-year period, which is very debatable. . . . When we had people talking about global warming, they didn't make their case at all. And . . . as far as I could see, they were shot down totally by the [skeptics] who were presenting the other side of the argument.

"I think any money that goes into this research is being taken from programs that really are proven to be necessity, programs that could be of great value to the American people. So I'd really oppose any extra money going into global warming research."

Jerry Mahlman's conclusions that global warming is a "harsh and inexorable reality," that "greenhouse warming is real . . . the state of the science is strong . . . it is a virtually certain bet that this problem will refuse to go away," didn't impress Dana Rohrabacher. Neither did the observation of a thirty-year climate scientist, Michael MacCracken, that "there's no doubt that" human activities are "dramatically modifying atmospheric composition and this is going to change the climate." Nor did the IPCC warning, cited by Robert Watson, that the planet is warming more rapidly than at any point in the past ten thousand years.

Nor, apparently, was Rohrabacher impressed by recent findings from the Goddard Institute for Space Studies that the half-degree rise in air temperatures since 1992 "may be a prelude to significantly higher temperatures" in the near future.

Understand that this is not politics as usual. This is something new.

The proof lies in the fact that research budgets for global monitoring programs bloomed during the 1980s. The largest increases in funding for scientific research into the global environment occurred during the administrations of Ronald Reagan and George Bush. In 1995 Congress turned that tradition of bipartisan support for science on its head. In its disregard of scientific facts and scientific method, it accomplished a violent attack on human knowledge.

In the summer of 1996, Rohrabacher's ominous collaboration with the oil and coal lobbies emerged to full public view—in the form of a vicious and intimidating assault on the integrity of a leading climate scientist.

The coal and oil lobbies found themselves unable to disprove the IPCC findings that global warming is under way, that it is attributable to coal and oil burning, and that the earth has entered a period of climate change. So they launched a second round of attacks—this time on the reputations and integrity of unsuspecting scientists who have been blindsided by the personal nature of the assaults.

In late May 1996, at a symposium in the Rayburn House Office Building, two of the leading IPCC scientists—Dr. Benjamin Santer and Dr. Tom M. L. Wigley—explained the findings contained in a new IPCC report, providing yet more evidence of human-induced global warming. Santer is a climate modeler at the Lawrence Livermore National Laboratory, while Wigley is a senior scientist at the National Center for Atmospheric Research.

The targeting of Santer was no coincidence. Barely a month later, Santer, Wigley, and eleven other researchers would detail their new findings in an article in *Nature. The New York Times* would call their findings "the most important . . . in a decade." An accompanying editorial in *Nature* would conclude that "the results of Santer et al.—using the available data and state-of-the-art climate models—provide the clearest evidence yet that humans may have affected the global climate." The Rayburn building session gave an advance look at the findings, as they were presented in the IPCC report.

In the Rayburn building, addressing a standing-room crowd, Santer used climate modeling to confirm that global warming cannot be attributed to the natural variability of the weather. He demonstrated that the measurable patterns of uneven warming coincide with patterns that would result from the atmospheric buildup of carbon dioxide accompanied by the distribution of airborne sulfates. The structure of that pattern is different from warming that is due to natural weather variations. In a follow-up presentation Wigley demonstrated how scientists have determined that the warming is due primarily to the release of carbon dioxide by human activities.

The session was attended by a number of industry representatives, including William O'Keefe, chairman of the Global Climate Coalition and an official of the American Petroleum Institute; Donald Pearlman, a Washington attorney who represents an undisclosed number of coal and oil producers and who quarterbacks the delegations of OPEC nations at international climate talks; and S. Fred Singer.

When the scientists had concluded their presentations, Pearlman and O'Keefe accused Santer of secretly altering the IPCC report of the previous year. They charged him with single-handedly suppressing expressions of dissent from other IPCC scientists. They derided Santer for eliminating references to scientific uncertainties.

These startling public accusations before a packed audience left Santer and Wigley visibly shaken. When Santer replied that one section of the 1995 IPCC report in question had been moved to another place simply for the sake of clarity, Pearlman scoffed. When Wigley pointed out that the chapter had been written by forty scientists and peer-reviewed by sixty others, Pearlman dismissed him out of hand.

Shortly thereafter, the coal and oil lobbies placed stories in *The Washington Times* and the trade paper, *Energy Daily*. They accused Santer of making "unauthorized" and "politically motivated" changes in the IPCC report. These stories, provided to the newspapers by the Global Climate Coalition, used incomplete and grossly misleading excerpts of the IPCC document that had been taken out of context. Playing on the recent "ethnic cleansing" atrocities in Bosnia, the public relations specialists of the GCC accused Santer of "scientific cleansing." The *Energy Daily* story, quoting extensively from a GCC handout, concluded: "Unless the management of the IPCC promptly undertakes to republish the printed versions of the underlying . . . report . . . the IPCC's credibility will have been lost."

Subsequently, in an op-ed piece in *The Wall Street Journal,* Frederick Seitz, director of the Marshall Institute, castigated Santer for allegedly excising references to scientific uncertainty. Wrote Seitz, "I have never witnessed a more disturbing corruption of the peer-review process than the events that led to this IPCC report." (Several months later, Seitz conceded the reports of his own Marshall Institute, which consistently deny any threat to the global climate, were not based on science but merely "represent opinion.") The story—with all its damning but unsubstantiated allegations—was picked up by *The New York Times.*

Predictably, the next issue of Pat Michaels's coal-funded *World Climate Report* accused Santer of raising "very serious questions about whether the IPCC has compromised, or even lost, its scientific integrity." The journal referred to the Global Climate Coalition as "a concerned group," never referencing the oil, coal, and automotive interests it represents.

Following this wave of disinformation, Santer wrote a letter that he sent to each of the authors of the 1995 IPCC report: "I am taking the unusual step of writing to you directly in order to keep you apprised of some very serious allegations that have been made recently by the Global Climate Coalition (GCC). . . . These allegations impugn my own scientific integrity, the integrity of the other Lead Authors of Chapter 8, and the integrity of the IPCC itself. I am troubled that this controversy has surfaced. I had hoped that any controversy regarding the 1995 IPCC Report would focus on the science itself, and not on the scientists. I guess I was being naive."

In an interview, Santer expressed his personal dismay at the unfair accusations. "All I want to do is to be done with this and get back to my science. But the last couple of weeks—both for me and my family—have been the most difficult of my entire professional career."

In their reply to *Energy Daily,* Santer and Wigley, as well as two other IPCC scientists (Dr. Tim Barnett, of the Scripps Institution of Oceanography, and Dr. Ebby Anyamba, of NASA's Goddard Space Flight Center), cited the "incorrect" nature of the material that had been provided to the paper by the GCC. The article's author, they added, "should at the very least have contacted one of the [IPCC authors] . . . in order to obtain a more balanced view of how and why revisions were made to this chapter." A separate response to *The Wall Street Journal* was signed by forty-two IPCC scientists.

Another letter to the *Journal* was signed by IPCC chairman Bert Bolin and by Sir John Houghton and Luiz Gylvan Meira Filho, co-chairs of the IPCC's working group on science. Santer's handling of the chapter in question, they wrote, adhered meticulously to proper IPCC procedure and violated no scientific ethics. "No one could have been more thorough and honest" than Santer, they added, in incorporating the final changes into the text.

Shortly thereafter, William O'Keefe continued the attack he had helped launch in the Rayburn building. He publicly called for "an independent review" to determine whether Santer had substantially altered the IPCC document. The attack prompted Santer to write his fellow IPCC scientists, "In effect, the Global Climate Coalition would like to put the IPCC—and my own scientific integrity—on trial."

Enter Dana Rohrabacher. When the coal and oil lobbies found themselves unable to discredit Santer, they enlisted the help of the "hear-no-science" congressman. In July 1996, responding to the lobbyists' urgings, Rohrabacher wrote to Secretary of Energy Hazel O'Leary, urging her to withdraw Energy Department funding from the laboratory that employs Benjamin Santer.

The year before, Representative George Brown, who had been Science Committee chair during the previous Democrat-controlled Congress, told an audience at Duke University that an extremist group of House Republicans were emotionally and ideologically inclined to ignore peer-reviewed science in favor of "fringe critics who . . . prefer to publish their criticisms in conservative . . . publications. . . . Those who are trying to understand how this new Congress is making environmental policy must understand that . . . [their motivation is one of] deep and fundamental belief. They contain the elements of a moral crusade that provides the believer with daily reaffirmation of meaning.

"Indeed, I cannot think of a single piece of new scientific evidence which would not be filtered out and rejected in the myth-making and soothsaying process. Whatever the findings, they will be suspect. They [will be seen in the new Congress] as the product of alarmist scientists who just want to keep their funding."

Concluded Brown: "The political dispute over policies to avert global warming is not fundamentally about science: it is about morality. It is about a vision of the future that requires collaboration and cooperation of all peoples to ensure our global security. It is about our obligation . . . to help preserve the global environment and to promote health, education, economic opportunity and freedom for everyone on the planet and for the generations that follow. . . . The time is ripe for . . . us to make the leap beyond narrow self-interest and understand a new reality: that we are all inextricably linked together on this planet, and that the welfare of the whole is of the highest importance."

On May 1, 1996, the same day the House Science Committee decided to cut the government's climate research programs, San Francisco broke a temperature record for this date when the thermometer at Mission Dolores hit 87 degrees at noon. The previous high of 86 degrees had been set 50 years before, according to the National Weather Service. Monterey also set a record that day, with a high temperature of 87 degrees. And in Salinas a 90-degree reading at the airport at 1:00 P.M. *tied a record set in 1947.*

Four days later, a succession of uncontrolled fires caused officials in Arkhangai, Mongolia, to declare a state of emergency. The fires, which had been burning since mid-April, destroyed more than 23 million acres of forest and rangeland, an area larger than the state of Maine. A snowless winter and unusually dry spring had fed the fires, which were then fueled by unseasonably intense and unpredictable winds. Chinese officials told reporters that more than 470 fires had struck Inner Mongolia that year, three times as many as in the same period the year before.

In Washington, D.C., the thermometer dropped to 32—an all-time low for May 14. Eight days later, it hit 96—an all-time high for that date.

FOUR

The Changing Climate of Business: Boom or Bankruptcy

UNTIL THE SUMMER OF 1996, THE DECEPTIONS OF BIG coal and big oil carried the day not only in Congress but in the international climate negotiations which were mandated by the United Nations in 1992. In the months leading up to the July 1996 round of climate talks in Geneva, however, the fossil fuel lobby finally began to encounter major opposition—from other segments of big business.

The fossil fuel lobby "wants you to believe that the science is divided, while business is united. In fact, the reverse is true," said Michael Marvin, director of one industry group, the Business Council for a Sustainable Energy Future. This group is seeking to jump-start the alternative energy business into an enterprise of worldwide scope, sparked by the climate threat. "There is no one in the business community—except for the oil and coal companies—who does not believe in the validity of the science and the reality of the climate threat," added an executive from the insurance industry, whose profitability is dramatically jeopardized by extreme weather events.

As these objections gather momentum, the oil and coal industries are becoming increasingly isolated from the larger business community. Even within the fossil fuel industry, some elements have separated themselves from the disinformation campaigns financed by coal and oil interests.

In a private meeting in Houston, executives of a branch of Royal Dutch/Shell decided that the company would not participate in the propaganda campaigns of Western Fuels and other industry giants because, in the words of one insider, "we didn't want to fall into the same trap as the tobacco companies who have become trapped in all their lies."

In a speech to the World Energy Congress in late 1995, John S. Jennings, chairman of a Shell subsidiary, said, "Some . . . organizations have, on occasion, shown that, through effective if unscrupulous use of the media they can influence public opinion significantly and with it the political mood to which democratic governments feel obliged to respond. I submit we all have an overriding interest in promoting calm, rational open debate on the whole range of energy issues. . . . But I believe it is imperative that the debate is based in principle on sound science, honesty and objectivity. . . . [W]e have to start to prepare for the orderly transition to new, renewable forms of energy and the lowest possible economic and environmental cost while sustaining secure supplies of conventional energy as the world economy hopefully continues to expand."

And in October 1996 BP America, a subsidiary of British Petroleum, along with the Arizona Public Service Company, a Phoenix-based utility, both announced they were withdrawing from the Global Climate Coalition. A BP official in London told a reporter for *Nature* that BP felt "its interests in the United States were not best represented by remaining in the coalition." An official of the Arizona utility explained his company's withdrawal from the GCC by saying that "Global climate change is a serious problem and we need to take steps to deal with it."

An early warning signal of the growing concerns of investors appeared in 1995, when London's Delphi Group, which advises large institutions on their investment policies, wrote a report recommending that banks, insurers, and other large institutional investors begin to withdraw their investments from oil and coal companies, despite their traditionally lucrative returns. Continuing disturbances in the global climate, the Delphi Group noted, could easily lead to high carbon taxes and enforced reductions on oil and coal use. "As a result," noted the report's author, Mark Mansley, a former financial analyst for Chase Manhattan Bank, "climate change presents major long term risks to the carbon fuel industry [which] has not been adequately discounted by the financial markets." The report recommended that investors "avoid maintaining long term overweight positions in the 'carbon fuel' industry" and added that "oil exploration companies are particularly vulnerable to climate change, since they are heavily . . . dependent on the existence in the future of buyers of oil assets at a good price." On the other hand, the report noted, "The alternative energy industry offers greater growth prospect than the carbon fuel industry. Diversification into this sector also offers substantial scope to offset the risks of climate change."

It is the world's insurers, as frontline casualties of climate change, who are leading the frontline opposition against the fossil fuel industry. Their executives are calling with increasing urgency for immediate and dramatic cuts in fossil fuel burning. In the last two years, they have found strong and growing support from large, multinational corporations in other sectors of the global economy.

The general manager of Swiss Re, the European insurance giant, recently took aim at the oil-and-coal-sponsored skeptics' claim that the increased weather extremes are merely due to natural climate variations. H. R. Kaufmann declared, "There is a significant body of scientific evidence indicating that the [recent] record insured loss from natural catastrophes was not a random occurrence. . . . Failure to act would leave the industry and its policyholders vulnerable to truly disastrous consequences." More succinctly, Franklin Nutter, president of the Reinsurance Association of America, said climate change "could bankrupt the industry."

In the international climate negotiations, the coal and oil

industries have aligned themselves, predictably, with the oil and coal-producing nations of OPEC. But the insurance industry has staked out the opposite position, aligning itself with a coalition of island nations who call for emissions reductions by the year 2010 to 20 percent below 1990 levels.

Insurance costs have been soaring as a result of weather extremes. As we have seen, since the beginning of the 1990s, a series of floods, hurricanes, and other severe storms have sent property insurance losses to unheard-of levels. In the 1980s insurance payouts for weather-related (nonearthquake) losses had averaged less than $2 billion a year, totaling $17 billion for the decade, according to the German reinsurance firm Munich Re. But from 1990 to 1995 that number skyrocketed to over $10 billion a year, with insurers paying $57 billion in loss claims from natural events in only five years.

Between 1990 and 1995, sixteen floods, hurricanes, and storms destroyed more than $130 billion in property—and caused deaths, homelessness, and psychological damage for the many victims of those catastrophes. Hurricane Andrew, in 1992, cost $30 billion, while the 1991 flooding in China cost $15 billion and the massive 1993 flood in the midwestern United States left more than $12 billion in damages.

On the eve of a March 1995 negotiating session in Berlin, an executive of the British insurance giant Lloyd's of London said his firm had asked several experts whether global warming was to blame for the unusually severe storms, droughts, and floods. "They told us, 'We can't prove there is global warming. But by the time we can, you chaps will be in real trouble,' " he said.

In 1995 fourteen of the world's largest insurance giants signed a Statement of Environmental Commitment pledging to incorporate environmental and, more specifically, climate considerations in their assessments. Meeting at the headquarters of the UN Environmental Programme (UNEP) in Nairobi, the group included, among other insurers, General Accident Fire and Life Assurance Corporation of Britain, Gerling-Konzern Globale of Germany, Sumitomo Marine and Fire Insurance Company of Japan, Swiss Re of Switzerland, and Uni-Storebrand SA of Norway.

In October 1996 one of the largest U.S. insurers, Employers

Reinsurance Corporation, a Kansas-based subsidiary of GE Capital, became the first U.S. insurer to sign the UNEP Insurance Initiative. In a letter to Vice President Al Gore, Kaj Ahlmann, CEO of Employers, noted that "From a personal and a corporate perspective, I find it of extreme importance."

"Insurers understand that environmental risks are business risks, pure and simple," says Hans Alders, European director of the UNEP. "They know that a few major disasters caused by extreme climate events . . . could literally bankrupt the industry in the next decade."

Still, as Franklin Nutter, of the American Reinsurance Association, warned in an interview last spring, "It is not easy to get people to look at the big picture proactively." Many industry decisions are based on relatively short-term considerations, he noted: "We need to get people looking at the longer-term implications." "All policyholders pay for these losses through higher insurance premiums," he added, "Ultimately, taxpayers also pay for the disaster assistance provided by government."

Underscoring the point, the Federal Emergency Management Agency (FEMA) has noted that between 1990 and 1994 the number of federally certified disasters jumped by 64 percent, costing taxpayers billions of dollars in relief operations. Nutter, citing projections by scientists that hurricane activity will increase substantially in the coming years, noted that the chances of a major hurricane striking a big city are great. "Insured losses alone from a direct hit could range from over $25 billion in New Orleans to nearly $55 billion in Miami."

In a direct shot at the antiscience, antigovernment congressional ideologues, Nutter noted that better building codes and improved risk analyses are not enough to deal with the economic magnitude of the climate threat. "A better understanding of weather patterns, natural climate variability, and fundamental shifts in climate—along with greater understanding of the potential impacts on society—are essential if we are to respond to threatening conditions in a cost-effective way. The research needed to build understanding in these areas constitutes the heart of the U.S. Global Change Research Program. This federally funded effort ultimately can help encourage better contingency planning, save billions of dollars in property losses, and most importantly, save lives."

Recently a third force—comprising a number of large multinational corporate players—has been positioning itself to take a leadership role in the climate debate. Calling itself the International Climate Change Partnership (ICCP), this group accepts the scientific evidence for climate change and agrees on the need for enforceable CO_2 reductions. It asserts that corporations—not governments—must design the global energy transition. Its membership, signaling a major corporate counterforce to the intransigent fossil fuel lobby, includes such heavyweights as Allied-Signal, AT&T, Dow Chemical, DuPont, Electrolux, Enron, General Electric, and 3M.

In 1996 another organization, the Business Council for a Sustainable Energy Future, announced that it, too, intends to be a prominent player in the negotiations. As we have seen, it aims to promote alternative energy interests. Smaller than the ICCP, the Business Council nevertheless includes some major corporations—Brooklyn Union Gas, Enron, Honeywell, Maytag, and several concerned utilities, as well as a host of alternative energy producers.

The evidence for climate change is making industry leaders increasingly sensitive to the implications of atmospheric warming for their own companies. Neither the press nor the environmental community has been able to prevent the fossil fuel industry from marginalizing the issue and stalemating progress in international negotiations. Perhaps these industrial associations of manufacturers and finance will succeed.

But make no mistake—the business opposition to the fossil fuel industries is driven first and foremost by considerations of the bottom line. The alternative energy industry sees an opportunity to emerge from its traditional status as a boutique industry into a major industrial player. The natural gas industry, with its relatively low CO_2 emissions, as well as the nuclear industry, see the climate issue as a way to increase their share of the world's energy markets. And the ICCP—that array of nonenergy industrial giants—sees the coming energy transition as a way to reinforce the dominance of transnational corporations over governments. They seek to devise and implement solutions to the climate threat that will not adversely affect their competitive standing.

Politics makes strange bedfellows, but climate change may be making even stranger ones. The leading environmental groups concerned about climate change—the Environmental Defense Fund, Greenpeace, the Natural Resources Defense Council, the Sierra Club, and the Union of Concerned Scientists—have been repeatedly frustrated in their efforts to get their message across to the American public. But in the ICCP they are finding support among some of the same corporate giants who have traditionally been their fiercest opponents.

The ICCP is a reincarnation of an earlier industry group of chemical users and manufacturers that had mobilized around the ozone-depletion issue. It crafted a successful, unprecedented partnership with the world's governments to phase out use of those chemicals that were damaging the atmospheric ozone layer. When the issue of climate change emerged, around the time of the UNCED Rio conference in 1992, the group reconstituted itself to address global warming.

The Montreal Protocol was heralded as an unprecedented collaboration between the governments and some of the largest transnational corporations. As an element of the new world economy, it has been largely ignored by most political and economic commentators. But the Protocol has implications for future corporate-government partnerships that may provide a strong platform for the ICCP in the coming rounds of climate negotiations.

If the fossil fuel lobby maintains that there is no scientific proof of climate change, the ICCP's executive director, Kevin Fay, has a very different perspective. "The fundamental science of global warming is pretty basic," Fay says. "There is some uncertainty about specific effects and impacts, but we understand that there is a long lag time for atmospheric greenhouse gases—and that it also takes a long time to develop remedies for the problem."

The ICCP's approach is more moderate than that of the insurance industry, with its call for immediate and drastic emissions reductions. But it is also far more progressive than many critics of industry would have expected: It agrees on the need for specific targets and timetables to reduce greenhouse gas emissions.

While the oil and coal lobbies take the position that all countries must adhere equally to the same cutbacks on fossil fuel use, the ICCP counters that the quickest way to scuttle any international agreement would be to impose that burden on such developing giants as China, India, Mexico, and Brazil. "We can't impose those limits equally on developing countries," Fay says. "At least not in the short term. That's a nonstarter. What we must do instead is invest in helping the big developing nations develop their own nonpolluting energy sources."

The ICCP "envisions a process," Fay continues, "that produces good information on what alternative energy technologies are out there, what developments are in the pipeline, and what the deployment of those energy sources will do to the global greenhouse gas profile. The next step we need to take is to assist those technologies to become competitive."

One key difference between the ICCP's approach and that of many environmentalists is that the ICCP insists that the coming transition to alternative energy sources must be designed by corporations rather than by governments. "We tell our members to bring their own corporate experience to the table in devising ways to manage the transition. But at the same time, we also tell them to step back and take a big look at the real climate issue," he says.

Industry should be held to a reasonable timetable in which to reduce emissions to mandated limits, the ICCP believes. But the business community, it contends, is far more able than government to achieve a smooth energy transition.

Governments might negotiate overly stringent reductions in too short a time frame, Fay argues, which would make their effort self-defeating. "If government sets an impossible task, industry's only response is to fight. . . . If the emissions targets are too low or the timetables too loose, then that will be a loser. The environment itself will see to that."

That is why the ICCP wants any solutions to be designed by the corporate community. If "we set a tough goal—but one that is reachable—then we have a good shot at success."

Still, off the record, many corporate activists voice a plea for strong political leadership—leadership from the highest levels of government—to effectively mobilize their corporate constituents. Fay puts that concern squarely on the record: "When you factor in the potential financial catastrophe to the coal and oil companies," he says, "as well as the stresses on the climate, it becomes very difficult to translate the situation into specific actions. I don't have one magic answer. But I know it begins with leadership. Leadership is our most critical need."

The landmark Montreal Protocol provided the ICCP with an impressive level of credibility. In accepting the climate threat and agreeing that any remedies be subject to external enforcement, the group clearly established common cause with much of the environmental lobby. Moreover, in its demand that any solution be designed and implemented by the corporate community, it is in a position to enlist the support of large segments of that community that are not involved with fossil fuels. Along with the insurance companies, then, the ICCP has positioned itself at the center of an extraordinary constellation of governments, environmental groups, and business interests.

The banking community, by contrast, has been one of the least visible segments of industry in the climate debate. In the short term, banks are essentially unaffected by climate disasters, since the mortgages they finance are virtually all covered by property insurance. Most of those mortgages, moreover, are normally sold by banks and thrifts to the secondary mortgage market, where agencies like Fannie Mae and Freddie Mac share the risks with the insurers.

But in the long term, the banking community too may feel the effects of climate change. At a 1995 conference on climate change for finance capital leaders, Sven Hansen, vice president of the Union Bank of Switzerland, noted that the ripple effects of climate change could adversely affect banks and their major customers. "Some of our clients are under major threat from climate change," he told the group. "In my opinion, it is the single most important environmental problem for the world today, but it does not threaten the survival of [the banking] industry generally as it does other industries.

Nevertheless, as banker to those industries and companies [most] threatened by climate change, we have to be concerned and we must recognize that the financial markets will be affected by climate change. . . . [S]ite contamination and lender liability are only the tip of a banker's 'environmental iceberg.' Climate change is, from my perspective, the mass lying underneath the water line. . . . [It] will probably surface soon."

Hansen noted some very specific ways that corporate clients of his bank could be affected by increased weather instability. "Dutch cities . . . like the Hague, Rotterdam, Amsterdam would be heavily affected by a one-meter rise in ocean level. . . . [C]ompanies like Philips and Shell, which have offices all over the Netherlands . . . could be considerably affected. . . . Now the link between this long-term scenario climate change [and] the drastic impact on our clients and finally on our bank should be clear."

Hilary Thompson of Britain's National Westminster Bank then noted some positive opportunities presented by the recognition of climate change. "I think," she said, "we have to balance the economic and environmental concerns so that we, as an industry, support businesses that are going to create sustainable wealth that will actually not only redress the historic harm that has [occurred] but also finance the research and development that is necessary to bring about change." To that end, Thompson noted that the economic implications of climate dictate banks' "funding new sectors, new growth markets and solar energy is one."

But the financial bottom line for the conferees was identified by Kaspar Mueller, a partner in the Swiss management and consulting firm Ellipson. "While wrong financial assumptions lead only to a revaluation of financial assets, wrong assumptions on the environmental future have much worse consequences," Mueller told the conference. "They lead to a definite crash. No change in price can close the ozone hole. That is the great difference."

The conference papers were compiled and edited into a book by Dr. Jeremy Leggett of Oxford University and Greenpeace UK. Leggett, who founded the Oxford Solar Investment Summit to promote the funding of clean energy technologies by large institutional centers of investment capital, sees an emerging "civil war in the energy industry." He believes that the financial growth opportunities of nonfossil-energy facilities are great, and cites as evidence the 1996 agreement by Amoco-Enron to construct a large-scale photovoltaic plant in India—using a system of panels that convert sunlight directly to electricity.

The major problem facing the renewable industry, Leggett explains, is that the markets are still too small to achieve the economies of scale necessary to compete with fossil fuels. As a result, the aim of the Oxford Solar Investment Summit is to create alliances among various segments of the capital and investment communities in order to build markets and create competitive investment opportunities in renewable energy. Without such a mechanism, the investment community won't adequately address the problem, he said in an interview. "But more and more, the banks and insurers are realizing that they can run but they can't hide, that the real solution lies on a worldwide switch to solar energy."

The Business Council for a Sustainable Energy Future, which grew out of the renewable energy industry, couldn't agree more. "The myth is that environmentally friendly energy sources will hurt the economy. But in fact, they will help the economy," says the council's director, Michael Marvin, formerly of the American Wind Energy Association. "The message we want to get to policy makers—both in the United States and abroad—is that a switch to a program of conservation, renewable energy, and natural gas is far more economical—and will help developing countries with their economic growth—in sustainable ways that save them money." He notes that the costs of a gas-fired cogeneration plant are far less than those of a coal-fired plant—producing less damage to the environment.

But Marvin stresses that for renewable energy and energy-efficient technologies to flourish, government help is needed. Before renewable energy can reach the scale its proponents envision, the industry needs reliable and continuing government policies to support their commercialization and deployment.

Without such consistency, the growth potential of the renewable energy industry will not be realized. As an example, Marvin noted that in 1992 the Bush administration launched a five-year energy policy, with bipartisan congressional support, that included, among other things, $50 million for research and commercialization of wind energy. But in 1995 Congress cut that funding by 80 percent—thanks in part to the efforts of Dana Rohrabacher, head of the same House subcommittee that voted to severely cut funding for global change monitoring programs. "Rohrabacher said he cut the funding for renewable energy in order to help cut the federal deficit," says Marvin. "But if you take away the very programs that make us competitive in order to reduce the deficit, what you have accomplished is actually quite negative."

The biggest danger to the U.S. economy, Marvin believes, is not the switch away from fossil fuels. The danger lies more in losing our competitive position as a leading producer of alternative energy technologies to other countries that are now making major investments in them. Wind turbines, for instance, already constitute Denmark's ninth largest export, while Japanese investment in solar technologies is soaring.

Marvin dismisses the assertions of oil and coal lobbyists that renewable energy sources are not yet sufficiently developed to meet the world's energy needs. "Most forms of renewable energy are already proven," he says. "But there is a ten-year window between the readiness of the technology itself and the readiness of the market to embrace it. Currently, utilities are skittish because they are still paying off debts from their earlier investments in nuclear plants. Couple that situation with the new deregulation of electric utilities, and there is a natural reluctance to make big investments in renewable energy sources until the industry sees how the new structure shakes down."

Noting that some photovoltaic facilities now produce electricity at 3.2 cents per kilowatt-hour—a price which is competitive with most utilities—and that large wind installations can produce it for 3 cents—Marvin says the public "should not fear leaving a coal-based economy. It is not nearly as hard as it might seem. It need not be that disruptive to the economy."

Kirk Brown of the Business Council explains that clean energy companies are already springing up all over the world—not only in the United States but in India, Germany, and Australia as well.

"These changes will happen anyway. The issue is how quickly," he observes. The critical point, he emphasizes, is that governments understand the economic benefits—not just the environmental imperatives—of switching from fossil fuels to renewable energy sources.

Clearly the use of alternative energy is growing. In the United States, despite the cuts in government support, Marvin says, the value of installed equipment that converts wind power to electricity has jumped from $600 million in 1993 to nearly $2 billion in 1996.

But what is most urgently needed, in the face of the gathering climate threat, is something that Congress views with horror—tax credits, subsidies, and other economic incentives to help proven nonpolluting energy technologies leapfrog full blown into the marketplace in such a way that they can compete with oil and coal. A production tax credit for wind energy, Brown notes, has already made it highly competitive. Today, in areas with strong and predictable wind currents, wind power is far cheaper than power derived from oil or coal. Were Congress to provide the renewable energy industry with the same $20 or so billion in tax credits and incentives that it provides to coal and oil producers, he says, those climate-friendly energy technologies could instantly become competitive with fossil fuels.

According to Alden Meyer, an energy specialist with the Union of Concerned Scientists, study after study has shown that every dollar spent on conservation, energy efficiency, and renewable energies creates far more jobs than the equivalent amount spent on fossil fuels.

It is a lesson that must be learned not only by governments but by the World Bank and the International Monetary Fund, whose investments fund many major energy projects in the developing world. The World Bank is a major "frustration," Brown says. Despite a handful of programs that underwrite the development of solar and other renewable energy systems in the poor nations, the bank is also providing billions of dollars for coal plant development in India, China, and other countries. Other observers note that under its new leadership, the bank is at last beginning to consider clean energy in its projects and is including energy conservation and renewable energy concerns in the loans that it leverages through private-sector banks. But the success of that new direction remains to be seen.

Meyer believes that for the energy transition to take place, investments must be lured away from oil and coal companies and put into renewable energy industries. A shift of only a few billion dollars a year could be enough to drive down the costs of renewables and make them competitive with fossil fuels, he argues. "That acceleration of renewable energy must happen soon," he explains, "if we are to avoid the next huge pulse of carbon from the developing world, as India and China and other countries bring more coal plants on line."

Every energy transition the world has undergone, Meyer emphasizes, has led not only to cleaner and more efficient forms of energy but to major increases in economic growth. "It happened when we changed from burning wood to coal, from coal to oil, and from oil to gas. If the next transition is handled properly, we would see a very big benefit to the world economy."

Meyer estimates that a near-absolute transition to a renewable energy economy could easily be accomplished at a cost of about $25 billion a year over the next ten years.

His estimate of what is needed to push the world into a new energy regime has one striking feature. According to a 1996 analysis by Doug Koplow of Industrial Economics, that number is actually on the order of $7 billion a year less than what the United States currently spends on subsidies for coal, oil, and nuclear energy. Koplow's study, which was commissioned by the Organization for Economic Cooperation and Development (OECD), found that nearly 90 percent of U.S. energy subsidies—the government's allocation of taxpayer-financed subsidies and incentives, which amounted to more than $32 billion in 1989—support the development of oil, coal, and nuclear energy; some $21 billion of that goes to fossil fuels. Using figures from 1989, the latest available at the time of the study, Koplow showed that only 10 percent—about $3.6 billion—went for conservation, energy efficiency, and renewable technologies.

In other words, shifting the current subsidies and tax incentives away from the oil, coal, and nuclear industries and making them available to solar, wind, and other alternatives could accomplish a near-total energy transition within a decade. The only remaining question is how much damage from the world's increasingly unstable climate we would have to absorb during the transition—and how long it would take planetary atmosphere to return to stability.

In the spring of 1996, the Kansas City–based Employers Reinsurance Corporation, the fourth largest reinsurer in the United States, hosted a conference on the impact of climate change. In an interview, Kathleen Raupp, an assistant vice-president, noted that her company—financially battered by weather-related disaster claims—is changing the way it does business. Employers Re has "already imposed caps on how much insurance we will write in certain areas." She explained that the company had recently identified certain flood-prone and wind-vulnerable areas in Texas, "hurricane alleys" in the Southeast, and other locations vulnerable to extreme weather events. "We have conducted a major reevaluation of our coverage," she said, noting that Employers Re is a subsidiary of GE Capital. "For instance, in storm-prone areas, we used to exclude areas along the coast—at most, say, ten miles inland. Now, in many cases, we are refusing to insure areas less than fifty miles inland."

That kind of change in the insurance industry is likely to exert a strong influence on other sectors of the business community. "I'm beginning to see banks, for instance, becoming much more concerned about lending on property that is vulnerable to these increasingly severe storms we're seeing," Raupp told me. But such movements in the financial community are only the beginning of what is required, she believes. "The fossil fuel industries are the biggest industries in the world. What can we do to make them change? We must get them to address the issue of emissions. We must provide a broader platform for the scientists and urge the scientific community to be more outspoken. Granted, we don't yet have absolute proof—a hundred percent certainty—about the impacts of climate change. But in my business, we make multimillion dollar decisions every day on far less certainty. Speaking as an insurance executive, my position is simply, if we can't afford the risk, we simply won't write the coverage."

Speaking not as an executive but as a human being, she added, "We might not be around in fifty years, but our children will be. That's the absolute bottom line."

But for the captains of the fossil fuel industry, the bottom line is maintaining their trillion-dollar-a-year commerce in oil and coal as long as they can—and supplementing those profits by selling energy-efficient and conservation technologies through their subsidiaries.

The fossil fuel industries are lobbying hard for a business-as-usual, free-trade response to the climate crisis that will help them penetrate and expand markets in the developing world without having to limit their U.S. greenhouse emissions. The idea is to turn the crisis into an economic "win-win" situation.

John Schlaes, executive director of the Global Climate Coalition, says that the only plan for an energy transition that his group would endorse would involve the gradual, voluntary development of free-market trade partnerships with developing countries. Under such partnerships, the energy giants would sell energy-efficient technologies to slow the growth of carbon emissions in Asia, Latin America, and Africa.

Any government-enforced solution, Schlaes warns, would raise the specter of an antigrowth bureaucracy. The people who want a mandatory reduction of oil and coal emissions, he says, are asking us "to change the world's relationships and superimpose a process on nations—something we have not successfully accomplished since World War II. The issues that this involves are staggering. They raise questions of international regulation, of national spheres of influence—even of national sovereignty."

Under the preferred scenario of the oil and coal industries, the large developing nations would reduce their emissions by using the energy conservation and efficiency technologies that they purchased from U.S. energy industries. The American energy companies, for their part, would gain new markets in the developing world. Through their new trading partnerships—for instance, U.S.-designed coal-burning plants could be sold to China. But at the same time the U.S. companies could continue to emit greenhouse gases at high levels because by selling clean technologies to the developing world, they would gain credit for the "emissions avoided" in those countries.

The level of poverty in the developing world, Schlaes cautions, means that meaningful trade partnerships will take years to develop. "To set up a network of partnerships such as we envision," he says, "requires major investments in the infrastructures of developing countries. It requires training their workers—which means a major investment in their educational systems. It means strengthening banking systems that we, in the West, take for granted. It means determining mutually advantageous markets. It means educating consumers in the developing world to new ways of operating. Most of all, it requires the development of trust between our industries and their governments. And that is best accomplished by a series of one-on-one business interactions."

Numerous studies—including some by the U.S. Department of Energy—indicate that nonfossil-based, nonpolluting energy technologies are already sufficiently developed to be deployed around the world. Schlaes dismisses those studies. "If we had proven, realistic, effective alternative energies, the marketplace would be promoting them," he argues. But this argument overlooks the fact that the current system of subsidies and tax credits for big oil and big coal keeps their products less expensive than those of their potential clean-energy competitors.

In the short term, the GCC's voluntary, market-based approach may be advantageous to U.S. business interests. But it ignores the urgency of the climate threat—and the potentially catastrophic disruptions that even a slight increase in the global temperature could trigger. "There are some people who want to dramatize the issue to push progress," Schlaes says, adding, "It will take time to devise proper strategic approaches that are not too costly to the U.S. economy."

Even as they oppose a United Nations–coordinated response

to the climate crisis, the GCC and other oil and coal industry activists appear to have developed their own new world order. For them, the global environment is merely a subset of the economy, and whatever nature sees fit to do will have to wait until it is subjected to a cost-benefit analysis to determine whether it will impose too harsh an economic burden on the U.S. economy.

And the fossil fuel lobbies are actively promoting this world-view. In July 1996 the U.S. delegation—which had hitherto refused to commit to binding emissions limits—changed its position and announced it would support the imposition of mandatory reductions in coal and oil emissions. The oil lobby proceeded to mobilize senators and congressmen from both parties—virtually all of whom represent oil- or coal-producing states or states with automobile manufacturers, trucking interests, and other industries which would be affected by a cap on fossil fuel burning. They warned the Clinton administration not to commit the United States to *any* mandatory limits without first spelling out for the public the negative economic impacts of such a move. "We do not believe the United States should enter into treaty commitments it cannot keep," said a letter to Clinton signed by Democratic Senators Bennett Johnston and John Breaux of Louisiana, Robert Byrd of West Virginia, Wendell Ford of Kentucky, Howell Heflin of Alabama, and Byron Dorgon of North Dakota.

With their relentless attacks on the world's scientific establishments and with their ceaseless interference in intergovernmental negotiations, the fossil fuel lobbies have been extraordinarily successful in blocking meaningful efforts to address the climate crisis. Perhaps the insurance industry and the other corporate players will be able to provide an effective counterforce to the coal and oil industries. Perhaps the world's diplomats will be able to overcome the opposition of the fossil fuel industries and approve emissions reduction targets—such targets could provide a new role for the insurance industry and the ICCP. If a group like the ICCP could bring to bear the efficiency and the experience of the corporate world to accomplish a global energy transition—without the traditional inertia and inefficiency of government bureaucracy—it could become a startling force for innovation and movement.

But I believe that a central pitfall would still remain. All the various business groups share a basic approach that would, if adopted, doom any long-term solution to the global climate crisis.

All of the corporate players in the climate debate—the renewable energy producers as well as the oil producers—see the climate crisis as another opportunity to sell yet another set of goods to the developing markets overseas. The coming energy transition, to them, is merely an occasion for exporting climate-friendly technologies to a worldwide market. If that vision prevails, I believe it will propel efforts to address the climate crisis toward inevitable failure.

Even the best-faith efforts of private industry would still constrain corporate groups to strike some balance between their competitive needs and the escalating instability of the global climate. Any such balancing act would subordinate the urgent requirements of the atmosphere to the requirements of the marketplace. If we wait until the developing world is able to afford to rewire all its homes and offices and factories, it will surely be too late to avert a slide into more tumultuous climate change.

What the climate crisis requires, I believe, is a plan that is not bound by the requirements of the marketplace. No climate plan that is designed to provide profits or protect corporate competitive advantage will work.

Even a nonmarket-based plan, however, would face two formidable obstacles. Scientists do not know what hidden thresholds lie ahead. They do not know what feedbacks will take effect, or when. They do not know at what point an unstable climate will become a cascade down a steep slope. They cannot yet predict whether or when the rate of warming will accelerate. So those who are trying to avert the crisis are left groping in the dark, forced to choose arbitrary emissions-reductions targets that are determined more by their political viability than by their correspondence to the actual climate situation.

Uncertainty about the rate of climate change, then, is one

unknown whose impact could scuttle a good-faith effort to reduce industrial emissions. The other unknown lies in the overwhelmingly complex and frustrating arena of international climate negotiations. Diplomats of these negotiations approach the bargaining table handicapped by often contradictory domestic and international political pressures. In their multilayered agendas, concerns about the global climate are pitted against the requirements of political alliances, security interests, domestic economic pressures, and all the other half-concealed forces that make the enterprise of diplomacy a phenomenally complicated balancing act. The knot of diplomatic agendas is so complex that untangling it may be beyond the capabilities of even the world's powerful corporations.

In early May 1996 in northern New Mexico, drought and hot, dry winds created the most volatile conditions ever seen there. Hundreds of firefighters converged to help combat the raging fires. "Conditions like this haven't ever been seen here—even in regular fire seasons," said Terri Wildermuth, a spokeswoman for the state Forestry Division. New Mexico and Arizona officials declared states of emergency. Forest fires have been rare in this high mountain country. Gary Schiff, of the Carson National Fire Service, said a fire that consumed a few hundred acres might happen once a decade. That spring, the fires consumed more than 142,000 acres. Fires in the southwestern United States result from seesaw weather patterns. Unusually wet years in 1993 and 1994 caused vegetation to flourish. But in 1995 the rainy season, which usually comes in July, arrived late. Then came a winter with little snowfall. By the time the annual spring winds whipped up in early April, the forests were so dry that a single spark could spread with a speed and fierceness that left firefighters dumbfounded. The normal fire season did not even begin until later in the year, officials said.

The same month a state of disaster was declared in the state of Queensland in southeastern Australia, where floods left one dead and three missing. Torrential rains flooded rivers in southern Queensland and northern New South Wales, cutting roads, leaving thousands of homes without power, and forcing scores of people to be evacuated, police said. The storms also whipped up huge seas that have pounded Queensland's southern beaches, known as the Gold Coast, leaving the beaches badly eroded. "At this stage we are looking at a million dollars to clean up the beaches alone. I imagine the storms have cost the tourist industry millions overall," Gold Coast mayor Ray Stevens said. A six-year-old boy died after slipping into a flooded waterway near a Brisbane golf course.

FIVE

After Rio: The Swamp of Diplomacy

IN JUNE 1992, 132 HEADS OF STATE ATTENDED THE UNITED Nations Conference on Environment and Development (UNCED) in Rio de Janeiro, where the issue of climate change was at the top of the agenda. The conference delegates approved the Framework Convention on Climate Change—a negotiating mechanism to promote agreement among all nations on how to respond to the gathering climate threat.

Since the conference, twenty-seven more governments ratified the climate change Convention bringing the total of signatory nations to 159. But numerous rounds of the talks established by the framework have yielded only incremental successes—punctuated by significant obstructions. Still, judged by traditional standards of international diplomacy, they have progressed apace—achieving slightly more than cynics might have expected. Judged, however, against the growing instability of the global climate, the talks have fallen far short of their original goals.

Much as the business community has responded to the climate threat with only those measures it considers economically permissible, so have the world's politicians restricted their own aims to what they believe is politically attainable. In my view, this calculation vastly underestimates the willingness of most people in the world to embark on a course of dramatic change to preserve a hospitable environment. But even the most ambitious political plans now on the table are dismally inadequate to meet the task of stabilizing the planet's atmosphere.

The Climate Convention talks have foundered for two specific reasons: the massive and confounding economic inequity between the wealthy nations and the major developing countries; and the exploitation of that inequity by the fossil fuel industry, in league with the world's oil- and coal-producing nations.

One bloc of countries—those most vulnerable to the effects of climate change—have taken the issue extremely seriously. But their concerns have been submerged by a tide of international jockeying, diplomatic posturing, conflicting domestic political agendas, and intense obstructionism by the oil-producing nations and their industrial allies.

In the fall of 1997 the convention delegates will meet again in Kyoto, Japan, to decide on targets and timetables for reducing greenhouse gas emissions. Yet as they prepare for the meeting, the initial mood of hopefulness four years ago has given way to one of disappointment. Observers who had initially hoped that governments would assume responsibility for the future of the planet now despair at the prospects of truly meaningful action.

"Even though the science is telling us that the consequences of climate change are just around the corner, I have real doubts about anyone's commitment to this issue," says one delegate from the developing world. "It takes more momentum among the developed nations than we've seen so far. But whenever it comes down to nailing someone to a position, it all seems to slip away. You see lots of nods and winks: 'After all, we're all politicians. We know that. Twenty years from now, no one will remember how we voted.' That partly explains why there is continuous posturing and very little else."

Of all the agendas in play at the various negotiating sessions, precious few reflect an undiluted concern for the stability of the global climate.

That the OPEC nations and their fossil fuel allies would play an obstructionist role in these negotiations was an expected complication. But their obstructionism has been made easier by a split between those delegations that truly want to address the problem and those that seek to maintain or increase their competitive advantage at the expense of any meaningful action. Some countries—Germany, Britain, and Denmark, for instance—have proposed significant emissions reductions, but their efforts have been blocked not only by the nations of OPEC but also by the United States, Canada, Australia, and New Zealand, who fear for the health of their domestic energy interests. Among the latter group, the oil and coal lobbies have found allies.

The implicit policy of many industrial countries seems to be to support any agreement that allows their corporations to profit from the climate crisis. If pushed, their fallback policy seems to be to accept any agreement that does not diminish their competitive position in the global economy.

From the outset of the negotiations, the strongest position in favor of tough standards was taken by a group of small island nations calling themselves AOSIS, the Alliance of Small Island States. But not even their position is based solely on altruism. Members of AOSIS such as the Philippines, Jamaica, the Marshall Islands, and Samoa fear that increases in severe hurricanes, intense rainstorms, and sea level elevation will flood them out of existence.

As Ambassador H. E. Tuiloma Neroni Slade of Samoa bravely declared: "the strongest human instinct is not greed—it is not sloth, it is not complacency—it is survival . . . and we will not allow some to barter our homelands, our people, and our culture for short-term economic interest." At the July 1996 round of talks in Geneva, Slade added: "It is absolutely necessary to act. We can no longer continue using the atmosphere as a dump for humankind's waste." As a result of such fears, AOSIS proposed one of the most stringent emissions standards for industrial nations—a reduction by 20 percent of their 1990 greenhouse gas emissions levels by the year 2005.

But even that standard, according to most calculations, is not sufficient to head off serious disruptions. A team of scientists from the Netherlands—which is highly vulnerable to the effects of rising sea levels—recently echoed the opinion of many climate scientists that in order to stabilize the atmosphere, the industrial world must cut its emissions by more than 50 percent below 1990 levels by the year 2010.

Jogged by those estimates, both Germany and Great Britain have proposed standards nearly as stringent as those proposed by AOSIS. They call for the industrialized countries to cut their emissions by 10 percent below 1990 levels by the year 2005 and 15 percent by the year 2020.

But these relatively bold initiatives have been dismissed by the United States, Australia, and OPEC as opportunistic and dishonest. For one thing, they charge Germany with playing to its domestic constituency, which includes the largest Green party in Europe. Moreover, those critics assert, Germany is uniquely able to sustain a large emissions cut by virtue of its reunification in 1989 with East Germany, which is far less industrial and emits far lower levels of CO_2. German emissions could therefore be reduced to meet the standard it proposes with little additional hardship. For its part, Britain, the critics argue, can readily support the emissions targets at no cost to consumers because in 1991 the United Kingdom decided for internal reasons to terminate its program of coal subsidies and to switch, instead, to far cheaper—and cleaner—North Sea natural gas.

The German and British proposals thus involve no sacrifice on their parts, according to their critics in the United States, Australia, and the Middle East. In fact, they charge, under an aggregated European Union (EU) cap contained in these proposals, the poorer countries of the EU would be permitted actually to increase their burning of fossil fuels. Ireland, Portugal, Spain, and Greece would be allowed relatively high emission margins so they can develop their own economies without cutting their fossil fuel consumption. And if, as anticipated, the EU expands to include a number of former

Communist nations, the combined cap for the enlarged EU would provide it with even more latitude to increase greenhouse emissions.

German officials counter that German citizens are bearing a significant tax burden to help finance an environmentally friendly reindustrialization of East Germany—a tax burden that would be unacceptable to Americans. And British officials say that whatever their initial motivations, in the last two years they have come to regard the climate crisis as very real and worthy of strong international action.

Britain's secretary of environment, John Gummer, went so far as to call for an emissions reduction target of 50 percent of 1990 levels in 1996. Gummer requested the imposition of a tax on airline fuel and called on governments to end all subsidies for oil and coal use. "This is an essential first step," he said, adding "there is no point in seeking to mitigate the effects of carbon dioxide while providing an inducement for people to use more."

In interviews with some American policymakers, it became clear that their criticisms of the German and British initiatives spring partly from a deep defensiveness about the United States' failure to move the negotiations much beyond stalemate. Yet some of those same policymakers feel a deep frustration with the constraints on the U.S. position imposed by the fossil fuel interests and their allies in Congress. As we have seen, in the summer of 1996 six Democratic senators from oil- and coal-producing states warned President Clinton not to commit the United States to any specific targets without first explaining to the public the economic consequences of that commitment.

The German and British initiatives have also evoked divided responses within Europe itself. Denmark, for instance, has already enacted a carbon tax as well as efficiency standards for its utilities in order to attain a 20 percent emissions reduction below 1990 levels by 2005. And the Netherlands has declared that it intends to stabilize its emissions at 1990 levels by 2000 and reduce them by 3 percent five years later. By contrast, Norway, with its lucrative North Sea oil reserves, wants no emissions limits at all.

Similar divisions appear throughout the developed world. While the Saudis and Kuwaitis hold out for no limitations on emissions—arguing, with the help of industry-funded skeptic scientists, that the science is still inconclusive—they have found strong allies in other fossil-fuel-producing countries. Australia, for instance, with its strong reliance on coal, argues that any set of emissions limitations must include a differentiation formula to accommodate the economic reliance of oil- and coal-producing nations on their fossil fuel exports. In May 1996 U.S. undersecretary of state Eileen Claussen warned Australia that it risked diplomatic abandonment by the United States if it did not give up its demand for special exemptions because of its large domestic coal reserves. The Australian officials refused to budge.

But divisions within the industrialized world pale before the yawning split between the wealthy nations and the poverty-stressed, less developed giants like India, China, Brazil, and Mexico. While the United States, Europe, and Japan can afford at least to give lip service to the high priority of the climate issue, the large developing countries, fighting to keep their economies above the undertow of grinding poverty, cannot. Yet until the economic inequity between the wealthy and poor countries is addressed with some degree of sincerity, no developing country will adhere to any agreement that restricts its economic growth.

Under the babble of diplomatic doubletalk in the climate negotiations, one clear message has emerged: The issue of global economic inequity is as critical as the carbon balance to the stability of the planet's atmosphere.

It is not a message the representatives of the wealthy world want to hear. Many Western diplomats, for instance, are quick to attribute the recalcitrance of countries like China and India to an attitude of indifference to climate change. Asked about China's commitment to the negotiations, one Western diplomat responded anonymously: "That's easy. They just don't care." And indeed, an environmental official of China's National People's Congress recently told *The New York Times:* "Two hundred years after the Industrial Revolution, the world economy has greatly advanced and the developed countries are the main beneficiary. About 80 percent of the world's pollution is caused by the developed countries and they should be responsible for those problems."

But by shifting the onus to the developing world, Western diplomats let their own wealthy countries of the North off the hook. The United States and its developed allies could exert leadership on climate change if they chose. The wealthy countries could take it upon themselves to address the strategically devastating global inequity in living standards. They could take steps to meet their obligation to future generations and ultimately fulfill their responsibility to the planet. Were they to make these choices, the rest of the world would clearly be willing to follow suit.

Dr. Anil Agarwal, director of India's Center for Science and the Environment, is at once sensitive to the climate issue and fiercely supportive of India's need for economic growth in order to support its population. "The United States is one of the leading nations of the world," Agarwal said in an interview. "When you provide leadership, it makes people proud of you. But if the United States continues its very high levels of oil and coal emissions, it will severely aggravate the already difficult position of vulnerable countries like Bangladesh and the Maldive Islands," both of which are highly susceptible to the ravages of rising sea levels. "That, of course, only propels unhappiness in the developing world against the United States. The United States has to appreciate what leadership is about. And, in my opinion, it still has some way to go.

"If you are looking for some kind of global peace and harmony within the limits of our planet, there has to be equity," he added. "If an agreement is reached that is not fundamentally equitable, it simply won't last." In other words, unless the developing countries are treated fairly in the imposition of any energy restrictions, unless their overriding issues of poverty and underdevelopment are addressed, they will sooner or later turn their backs on any international agreement.

Under the terms of Climate Convention, restrictions on coal and oil emissions would initially be imposed only on the industrialized countries. But in subsequent stages, as diplomats envision them, similar restrictions would be placed on poor countries. With such restrictions, the developing world will be unable to continue its current, potentially catastrophic, and growing rate of fossil fuel burning. But by delaying the restrictions on these emissions, the

Climate Convention framework acknowledges that, for the time being, the developing countries must give priority to creating jobs, housing, schools, health care facilities, and infrastructure.

Still, even with the weight of overwhelming poverty, the developing giants are acutely aware of the planetary limits to fossil fuel burning. As Agarwal, speaking for legions of experts in the developing world, points out: "The ultimate resolution of the problem is the worldwide use of appropriate energy technologies. But until there is a fair and effective agreement, and the United States and Europe and Japan take the first steps, the developing countries have no reason to take any action. It is only when the United States and Europe and Japan take the lead that the arguments about the global climate will have some moral weight behind them."

More than moral weight will be required of the wealthy nations. A transfer of wealth—in the form of clean energy technologies—will be necessary to help the poorer countries leapfrog over the archaic and destructive type of industrialization that is powered by coal and oil and use energy from the sun, the wind, and the rivers to develop their economies.

Without such assistance, a doubling—and perhaps a tripling—of atmospheric carbon dioxide concentrations is probably inevitable. Today, for instance, China is staggering under the pressure of an increasingly fragile food supply and diminishing water resources. At this point the government sees no alternative but to promote the country's economic growth as rapidly as possible. Thus, while energy consumption in the United States, Europe, and Japan rose by about 28 percent between 1970 and 1990, it rose by almost ten times that amount—208 percent—during the same period in China. Under current estimates, moreover, Chinese coal consumption—which equaled that of the United States in 1990—will be more than twice U.S. consumption ten years from now.

As large as it is, China is only one of many sources of the emerging boom in Asian greenhouse gas emissions. In May 1996 the environment ministry of Japan estimated that emissions of carbon dioxide and other greenhouse gases from fifteen Asian nations will more than double in the next thirty years—a projected increase of 150 percent above 1990 levels by the year 2025. Using data from other Asian countries as well as from the World Bank, the Japanese ministry reported that Asian greenhouse emissions will account for 36 percent of the world's emissions by the year 2025 and for 50 percent by the end of the next century, according to a story in Tokyo's *Daily Yomiuri.*

The situation in China is so serious that it is sparking alarm even among its own researchers. Oceanographers at China's State Oceanic Administration estimate that if current trends continue, rising sea levels will trigger massive flooding. "Sea levels will rise up to three feet during the twenty-first century," says Du Bilan, a researcher with the administration. Unless a massive program is undertaken to build coastal-protection bulwarks and seawalls, the researchers project that economic losses from a twelve-inch rise in sea level will reach $1.63 billion. "Since coastal areas are home to about half of China's cities and 40 percent of the country's total population of 1.2 billion, the government and society should attach great importance to monitoring sea level changes," Du says. Because surging waters would increase the salinity of farmland as well as cause coastal erosion, Du's group proposes rebuilding existing dikes and dams while launching a massive construction program for coastal protection in the near future.

In an official 1996 white paper, China's cabinet-level State Council appealed to the developed nations to help China lessen its reliance on fossil fuels, since it was escalating both the levels of air pollution and the incidence of lung cancer. Noting that five of China's largest cities are among the ten most polluted in the world, the white paper said that "the cities with concentrated industries and populations suffer from serious air pollution. Acid rain has occurred, and the situation has gone from bad to worse." Moreover, according to the Chinese environmental agency, the death rate from lung cancer increased from 1994 to 1995 in rural areas, where 80 percent of China's 1.2 billion people live. Indeed, in those areas respiratory diseases have become the leading cause of death. China is under no specific obligation to limit its emissions of carbon dioxide, the agency said, but it voluntarily pledged to reduce its emissions, by the year 2000, to 1995 levels, out of "responsibility for protecting the global climate."

There are other signs—small but hopeful. They show that the people who inhabit the poor nations are far more sensitive to the requirements of nature than the diplomats of the oil- and coal-producing nations—including the United States—would have us believe.

Even without the pressure of international agreements, several developing countries are already taking steps to change their energy sources. By the end of 1995, for instance, after changes had been made in its tax code, India had installed facilities that would produce some 900 megawatts of wind-generated electricity. It also announced plans to generate another 500 megawatts of solar (photovoltaic) electricity, as well as additional power from small hydroelectric facilities and from sixty new small-scale biomass generating plants. Over the last five years, India has committed $400 million to solar power installations—including a joint venture with the Enron and Amoco corporations for a 50-megawatt solar plant that is expected to provide electricity to 200,000 homes. Singapore, the Philippines, Malaysia, Thailand, and Indonesia have all cut their energy use by more than 20 percent over the last decade, by implementing energy-efficient building codes. And the Marshall Islands, lacking indigenous fossil fuel resources, is installing a thousand solar hot-water systems in order to wean its 55,000 residents off imported fuel oil.

In fact, some developing countries are now exporting renewable energies to other developing countries. India recently donated a dozen solar panels and streetlights to Cuba, where they were installed in the village of Magdalena, according to the *Los Angeles Times.* With help from China, Cuba has also built 206 small hydroelectric power plants. As small as these steps are, they nonetheless speak to the receptiveness of developing countries to renewable energy sources—despite the massive problems they face—because of their economic and environmental benefits.

Even while developing countries are sensitive to the global environment, however, they are obliged to support complex networks of political relationships. The Philippines, for example, shares interests with both the large developing countries and with AOSIS, of which it is a member. Antonio Gabriel LaVina, an official in the Philippines Ministry of the Environment, explains that the Philippines are worried about environmental devastation.

"Certainly we have strong economic priorities," he says. "But because of the potential impacts on us of rising seas and increased storms, we want commitments from the developed countries to reduce their emissions. We ourselves actually see the climate change threat as an opportunity to review our own energy systems and become less dependent on fossil fuels. If international policies are negotiated that require us to change our energy system, we don't think that would be at all bad. There is great opportunity here to reduce our own dependence on coal and oil."

But the Philippines are still obliged to continue fossil fuel burning until the developed countries take leadership by reducing their own emissions. "We won't be able unilaterally to accept stringent limitations on our fossil fuel use," LaVina explains. "The increase in energy demand in the Philippines is very strong."

All nations should take effective action to reduce fossil fuel burning, LaVina emphasizes. "At bottom, we should all reduce emissions, not just the United States and Europe, but everywhere. . . . Any kind of action leading to less use of fossil fuels is critical to the global climate," he said, noting the Philippines are vulnerable to ocean rise and severe storms. But, he added, developing countries "have every right to insist on equity. The historical argument is very powerful and correct."

Despite its vulnerability to sea level rise and severe storms, however, the Philippines have aligned themselves with the large developing nations because of their common problems. "China and India want a [looser] timetable," LaVina explains. "We have supported them for reasons of equity, and it is a sound thing to do." The country also has a close trade relationship with Saudi Arabia, which has employed substantial numbers of Filipino workers on its many construction projects. This relationship must be preserved, despite the stresses of the climate change issue. "The reality is that when you make decisions, you consider all your national interests," he says. "We have important relations with other countries. And even though our primary interest in this issue is environmental, there are very legitimate issues of our national interest that prompt us to retain our relationships with Saudi Arabia and the others."

The massive difference in economic pressures and priorities between North and South has not only produced diplomatic deadlock. It has provided a most useful wedge for those who most want the negotiations to fail.

Throughout the ongoing Climate Convention negotiations, the OPEC nations and their industry counterparts have consistently warned the large developing giants—China, India, Brazil, and Mexico, for instance—that the "climate scare" is based on flawed science and is basically a plot by the wealthy countries to keep them relatively poor. Consequently, they have urged those countries to accept nothing but the loosest future restrictions. At the same time, the OPEC nations and the fossil fuel industry representatives have put the United Nations on notice that they will accept no restrictions that do not fall equally heavily on the developing giants. Anything less, they argue, would be fundamentally unfair to fossil fuel producers, whose income depends on sales of oil and coal.

It is a strategy designed to guarantee the failure of the talks.

Clearly any attempt to impose the same restrictions on the poor countries—whose per-capita consumption of coal and oil has been comparatively tiny—as would fall on the rich countries—who have built their dominant industrial wealth on a base of fossil fuels—amounts to nothing less than "environmental colonialism," according to Sunita Narain, an Indian researcher and co-author of *Global Warming in an Unequal World.*

But that approach is what the fossil fuel lobbyists are pitching. Saudi Arabia and Kuwait, in league with representatives of the U.S. oil and coal industry, have successfully used these tactics to keep the negotiators from moving forward.

At a preliminary negotiating session in New York in February 1995, China and India, both of which have vast coal resources, argued that unless the United States leads the way by significantly cutting its own emissions, their obligation to develop their own economies outweighs their obligation to pressure the global environment. The range of initial positions, according to observers, left some room for negotiation. But that hope was scotched by the OPEC bloc, which steadfastly opposed even the vaguest of goals. The delegations from Saudi Arabia and Kuwait were especially intransigent, according to observers, who express particular irritation with the role of Donald Pearlman, a former official in the Reagan and Bush administrations.

Pearlman, who is a partner in the Washington law and lobbying firm of Patton, Boggs and Blow, has basically functioned behind the scenes as the parliamentary quarterback for the OPEC delegations, directing their efforts to hamstring the negotiations. On several occasions Pearlman was seen passing handwritten notes to Saudi and Kuwaiti delegates, advising them to oppose specific treaty language and providing them with alternative wording more favorable to their interests. Several observers witnessed OPEC delegates carrying Pearlman's notes into negotiating sessions and consulting them during the deliberations.

In the end, the OPEC delegations prevailed in that February 1995 meeting. Supported by the United States, Japan, Australia, Canada, and New Zealand, the meeting rejected calls to limit emissions, declaring such action premature. The following month, a subsequent negotiating round ended with an agreement to negotiate a vehicle for reducing emissions to be adopted in Japan in 1997.

Under the Foreign Agents' Registration Act, Pearlman is registered as a representative of Abu Dhabi, Oman, Quatar, the United Arab Emirates, and other oil-producing countries. But in a telephone interview he emphatically denied that he has ever formally represented any foreign government in connection with international climate deliberations. He did not deny counseling OPEC delegates.

Several years ago Pearlman founded a nonprofit organization called the Climate Council. According to its IRS charter, its purpose is to represent U.S.-based energy companies "whose business could be adversely affected by laws relating to potential global climate change." Pearlman secured official Non-governmental Organization (NGO) status for his Climate Council, which gives him access to briefings from the official U.S. delegation—a situation that enrages his critics in the environmental community, who have dubbed him King of the Carbon Club. Environmentalists charge that Pearlman's official nonprofit status, together with his ties to the Saudis, Kuwaitis, and other OPEC nations, allow him to be privy to information from both camps in the negotiations. Jennifer Morgan, coordinator of the U.S. Climate Action Network, an affiliation of U.S.-based environmental NGOs, says that given Pearlman's relationship with OPEC delegations, he should not be permitted to "enjoy official nonpartisan, nongovernmental status." Pearlman dismisses that criticism, insisting he has never received any compensation from any foreign government for any work related to climate issues.

Like the skeptic scientists and their ideological supporters, Pearlman has also criticized the work of the IPCC. In an interview, he bemoaned the politicization of the IPCC process, saying it was being contaminated by the political agendas of its member nations. "The IPCC has been heavily politicized from the beginning," he said. "It is as true of highly technical scientific issues as it is of economic ones. One can count on the fingers of one hand the times when any nation has taken a position before the IPCC . . . that would not further its policy preferences."

At the negotiating session in March 1996, the level of discord was as high as ever. As the delegates began to consider several agenda items, they were interrupted by delegates from Saudi Arabia, Kuwait, and China, who claimed they had not had time to read the IPCC report—although they had received them three months earlier, at a meeting in Berlin. But even if they had received the reports only the day before, that excuse would still lack any credibility. According to a newsletter of the environmental NGO community: "This blatant, transparent attempt to undermine the key conclusions of the [report] was even more unbelievable because these same countries negotiated every line, word by word, in the final months leading to the [report's] approval . . . last December."

The March 1996 negotiating sessions considered the question of international energy efficiency standards for appliances—a fairly minimal but nonetheless useful step toward emissions reductions. Initially, most of the delegates were agreeable to implementing the relatively innocuous efficiency standards. But following discussions with oil and coal industry representatives, a number of delegates from developing countries rejected the standards, on the ground that they constituted a nontariff trade barrier and hence violated the General Agreement on Tariffs and Trade guidelines. All these delegates used the exact same wording, in which they had been coached by the lobbyists. Not even the adoption of minimal efficiency standards could slip by the guardians of the oil and coal interests.

At that same negotiating session one island-nation delegate suggested that, rather than have 150 governments negotiate regulations for the auto industry, representatives of the leading auto manufacturers might meet and discuss ways in which they could cooperate in setting global standards and developing less-polluting vehicles.

The industry response was disheartening. "Business representatives present at this meeting greeted the suggestion with loud, dismissive and cynical laughter," according to an account in *ECO,* the environmental NGO newsletter. The account concluded: "Most . . . of the business NGOs now straining the capacity of the observers' gallery come to Geneva to slow the Convention's advancement. Their most memorable contribution to last week's workshop . . . for example, was to ask, with naive cynicism, whether it might not be more cost-effective to build one meter sea walls around the globe than to invest in mitigating climate change."

As for the OPEC delegations, *ECO* noted that "Saudi Arabia is still trying to argue that producers of polluting fossil fuels have a right to a special representation . . . in the Convention process. But as far as its core work goes, little interest has been shown, and many [OPEC delegations] have arrived without evidence of prior preparation."

The sustained campaign of diplomatic sabotage continues with the relentless obstructions that are raised when IPCC scientists try to get their message out.

Dr. Kevin Trenberth was the head of an IPCC scientific group that, in late 1995, wrote a policymakers' summary of the IPCC scientific draft report. Trenberth, the lead author of the summary, said it was in fairly good shape going into a negotiating session in

Madrid near the end of November 1995. But he noted with great frustration in a letter after the session that "there were deliberate attempts to obfuscate and undermine the documents by the OPEC nations, principally Saudi Arabia and Kuwait. . . . Even with an overwhelming 93 to 2 vote in favor of particular wording, proceedings were delayed time and time again seeking alternative language that would satisfy everyone. Sidebar meetings were held by groups of 10 or so which were supposed to resolve wording of particular sections. But neither the Saudis nor Kuwait would participate in these and instead offered continual amendments. . . .

"In the final throes of trying to get a document of some sort, the Saudis continued with every ploy including saying there was no quorum," despite the lack of any quorum requirement. "Part of the strategy seemed to be to avoid having a document at all," Trenberth surmised, so frustrated by the blatant obstructionism that he considered taking his name off the document.

Taking their cues from Donald Pearlman, the Saudi and Kuwaiti delegates objected to virtually every word of the summary. Noted Trenberth: "All of the [resulting] wording was that of the Saudis in spite of overwhelming support for alternatives. Every word was approved separately. The result was no macro perspective of the whole document or even sections. It was, at best, disappointing."

The obstructionism was epitomized by the fight over the heading of a section on sulfate aerosols. After intense negotiations, the section was finally titled "Anthropogenic Aerosols Tend to Produce Negative Radiative Forcings," which, Trenberth said, "is true but says nothing. Most people favored my suggested wording, which was 'Anthropogenic Aerosols Tend to Mask Global Warming.' " But that title was rejected by the OPEC delegates. "All this does not augur well for the future of the IPCC. I believe it cannot continue unless substantial changes in rules of procedure are made," Trenberth added.

In a subsequent letter to another IPCC official, Trenberth asked: "The question . . . [is] how to deal with small groups whose real intent is to subvert the process? Whether or not this was really the purpose of the Saudis and Iran, this question remains an important one and I believe that procedures should be put in place that are understood by all participants as to how these things should be dealt with in future."

The oil and coal interests have made the prospect of any meaningful emissions reduction in Japan in late 1997 highly doubtful. Like their U.S. counterparts, the OPEC delegates continue to insist that the state of the science is too ambiguous to take substantial action. In November 1995, after the Saudis and Kuwaitis worked so hard to thwart IPCC scientists in Madrid, Merylyn McKensey Hedger, an officer for the Worldwide Fund for Nature, told a writer for *Nature* magazine that "the situation is too serious for countries such as Saudi Arabia and Kuwait to continue . . . to subvert the IPCC process." But that complaint was rejected out of hand by Mohammed Al-Sabban, head of the Saudi delegation. "Saudi Arabia's oil income amounts to 96 percent of our total exports," he said, adding: "Until there is clearer evidence of human involvement in climate change, we will not agree to what amounts to a tax on oil."

In Geneva in the summer of 1996, eleven oil- and coal-producing nations continued to attack the findings of the scientists, voting to reject the IPCC's scientific report on growing climatic instability. Kuwait, Saudi Arabia, Syria, Iran, China, Nigeria, and five other oil-producing states declined to approve the IPCC text without substantial modification to its language.

In response, Bert Bolin, then chair of the IPCC, declared: "This is the work of two thousand scientists, reviewed, revised and reviewed again. There is no compromise in any direction. Do not trust any individual scientists, not even me; look at the work of all these, a balanced view." Referring to this and other recent attacks on the IPCC, Bolin said, "We reject accusations and allegations made against us. We have completely and carefully done our work. It is the best science on climate the world has to offer. I stake my reputation and honor that it is so."

Unfortunately, Bolin's honor means little to the fossil fuel industries, whose lobbies will do whatever they can to undermine any process—scientific, diplomatic, or regulatory—that could lead to enforced limits on coal and oil burning.

At the Geneva meeting U.S. undersecretary of state Timothy

Wirth heralded a new, more aggressive U.S. policy when he announced for the first time that the United States would support a binding international agreement to limit emissions. "Saying that we want to have a target that is binding is a clear indication that the United States is very serious about taking steps and leading the rest of the world," Wirth said. "We will be getting more and more specific as the next year and a half rolls through. . . . We are not interested in grand rhetorical goals that are impossible to realize. We want the negotiations to focus on outcomes that are real and achievable."

Wirth called his announcement "a big deal"—and on the scale of diplomacy, perhaps it was. But if, after four years of talks, a call for mandatory but unspecified emissions targets is a "big deal," then progress is being measured according to standards of diplomacy, not those of the biosphere.

The actual targets the United States envisions, Wirth conceded, are far less than the 20 percent reductions proposed by AOSIS, Britain, and Germany. "Our preliminary analysis of some of these targets," he explained, "suggests that [they] are neither realistic nor achievable—either because they would compromise other important principles, such as the need for flexibility in . . . of implementation, or because they involve timeframes and objectives that are not consistent with national and international prosperity."

But even this minimal hardening of the U.S. position—an agreement to specify mandatory emissions limits in the future—generated an immediate counterattack from the oil and coal lobby. John Schlaes, executive director of the Global Climate Coalition, declared in Geneva: "The Clinton administration is willing to risk a wild ride for Americans on a roller coaster economy while giving developing countries a free ride. It is clear the administration is pursuing a political agenda. This decision was made without adequate economic facts and is based on uncertain climate science." Schlaes accused the administration of proposing a plan that "could eliminate millions of American jobs, reduce the nation's ability to compete globally, and force Americans into second-class lifestyles."

Still, the U.S. change of position did produce the first cracks in the hitherto unified front of the fossil fuel lobby. An official of the Edison Electric Institute (EEI), which represents six hundred electrical utilities, refused to endorse the GCC position. Instead, the EEI's Charles Linderman said, "We sell electricity and we do not care where it comes from. The power lines do not know [whether] it comes from coal, wind or solar cells. We are with the future. We know we cannot go back to the old days."

How the oil and coal lobbies will respond, in the final rounds of negotiations, to what they see as a major betrayal by the United States remains to be seen.

Adamantly resisting any enforceable emissions limits at all, the GCC has steadfastly argued for the gradual, voluntary, and market-driven development of nonpolluting energy systems in the developing world. According to a GCC-commissioned study, Schlaes said in an interview, the AOSIS emissions standards, if adopted, would cost the United States 3.5 percent of its gross domestic product. "We simply cannot afford that. It's as simple as that," Schlaes said recently.

But a group of environmental NGOs has estimated the total global cost of stabilizing emissions at a far lower price: $80 billion per year by 2020. "At an average cost of about $15 per person per year, this is a very small price to pay to take the first step towards avoiding the dangerous impacts of climate change predicted by the IPCC," according to the group, which notes that estimates like the GCC's "do not compare [those] costs with the economic, social and environmental costs of the impacts of climate change if insufficient action is taken."

Nevertheless, the GCC, which represents U.S. auto manufacturers as well as coal and oil interests, insists doggedly that emissions caps would simply be too costly to the United States. Along with the oil and coal interests, they continue to push for successively weaker measures, to be adopted at the Conference of the Parties in the fall of 1997.

The spring 1995 meeting in Berlin determined that specific emissions cuts and a timetable would be adopted in 1997 for the world's industrial nations. But OPEC and its corporate allies want only the gradual development of voluntary trade partnerships with developing countries, which would promote the sale and distribution of more efficient energy sources. "It will take a long time for the developing world to evaluate their own needs and relationships before we can talk about technological changes," says Schlaes. "Meanwhile, since U.S. coal plants are two to five times more efficient and cleaner than the Chinese designed-generating facilities, let's sell them our plants while we figure out the real threshold for carbon dioxide concentrations. The more efficient plants are good for the environment and also good for U.S. business."

In 1992 the delegates to the Rio conference pledged to alleviate the most ominous environmental problem ever to confront humanity. Four years later, the grand rhetoric has given way to considerations of competitive advantage and cost-effectiveness. In particular, the relentless pressure from big oil and big coal has led to a striking retreat from international leadership on the part of the United States—the only country that can mobilize a consensus for action, given the welter of diplomatic agendas at play. The result has been the puny and inadequate measures now on the bargaining table.

The fossil fuel interests have either weakened or abandoned several sets of policies that have been proposed. Most notably, they have to date resisted embracing any specific emissions reductions targets and timetables. Even though the United States pledged to support some undefined targets and timetables in Geneva, it is clear that whatever numbers are finally decided will be far too small to bring stability to the global climate.

The second type of measure that the industry has prohibited is the carbon tax. Carbon taxes have long been accepted in Western Europe—both to discourage excess energy consumption and to finance environmentally beneficial programs. But the Clinton administration, bowing to the country's current antitax sentiment, has thrown in the towel on this idea—without even attempting to make a compelling public case for it.

One plan that the United States apparently will promote involves a scheme of "emissions trading rights," similar to the "pollution permits" plan instituted in the United States to reduce acid rain. Under an "emissions rights" trading program, an aggregated total of oil and coal emissions would be allocated among the various countries of the world. If an industrialized country were to exceed its own emissions quota, it could purchase unused rights from lower-emitting countries—thus rewarding the less-polluting nations for not contributing to the buildup of greenhouse gases and providing them with capital to develop renewable sources of energy. Conceptually, that kind of plan could redress some of the inequity between the countries that produce the most emissions and those who produce the least.

But the emissions-trading-rights scheme gets bad reviews from both fossil fuel interests and from environmentalists. The fossil fuel community dismisses the scheme as too bureaucratic. And most environmentalists believe it falls far short of the measures that are actually needed to address the increasingly unstable climate.

These objections aside, the fundamental problem with the scheme involves an issue of potentially profound contention: how to allocate the emissions rights in the first place. The developed nations would want the allocations to be derived from their own previous baselines—that is, their actual emissions in, say, 1990. Many coal- and oil-producing countries, moreover, have insisted that their allocations must compensate them for the higher costs they would incur by reducing their fossil fuel production.

By contrast, the poorer countries would undoubtedly insist on higher allocations for their own countries, given their massive need for development and the fact that they have historically contributed only a small fraction of the world's greenhouse gases. A number of poor countries have signaled their insistence on a per-capita allocation system. But such a system would give countries like India and China huge allocations relative to the wealthy nations because of their enormous populations.

Ultimately, any emissions-trading-rights plan seems doomed to endless haggling over allocation of emissions-rights quotas. The problem cuts directly to the central conundrum: the massive economic inequity between the world's minority of wealthy nations and its majority of poor ones.

As a fallback proposal, U.S. officials are touting a system called "joint implementations," which Wirth has vowed will be linked to specific emissions reduction targets. Joint implementation would permit heavily polluting countries like the United States to spend money—not necessarily to cut its own emissions but to fund renewable energy systems in developing countries. The resulting credit for "avoided emissions" would then be negotiated between the buyer and seller countries, as part of the price. ("JIs," as they are called, need not involve alternative energy. The reindustrializing economies of Eastern Europe, for instance, are potentially large markets for energy conservation and efficiency technologies, while Costa Rica and Venezuela are particularly receptive to tree planting and forest conservation as ways to enhance the biosphere's capacity to absorb carbon dioxide.)

Joint implementation would most likely play out as a scheme by which industrialized nations could buy their way out of domestic emissions reduction requirements by gaining credits for "emissions avoided" from the developing countries to which they sold their wares. It comes as no surprise, then, that absent a strong and enforceable emissions cap, these joint implementations are the market-based plan of choice of the oil and coal industries. But even if the negotiators should agree to a mandatory cap, joint implementation would fall far short of meeting the requirements of the earth's atmosphere.

Despite President Clinton's voluntary program to reduce emissions—and others like it in the industrialized world—1995 saw the burning of a record high 6.1 billion tons of fossil fuels, according to the Worldwatch Institute. United States carbon emissions continued to rise in 1995, according to a report by the Department of Energy, which attributed the rise to a lack of implementation of energy efficiencies. In the summer of 1996 the EPA sponsored a Climate Change Analysis Workshop, where environmental analysts Kilaparti Ramakrishna and Andrew Deutz of the Woods Hole Research Center addressed this situation.

"Given the appalling state of compliance with the modest commitments . . . by the industrialized countries and their reluctance to advance a serious political commitment in favor of a strong [set of emissions targets and timetables]," the analysts warned that "the feeling is gaining ground that the industrial countries are not serious about the problem of climate change and that the developing countries therefore need not put in place . . . any major policies that will over time reduce their carbon contributions to the atmosphere." Yet it is urgent that they do. "A number of major investment decisions that will tie these [developing] countries to ever increasing greenhouse gas emissions are being taken now. . . . All this places a special burden on the industrialized countries to agree on a cohesive strategy to reduce their emissions . . . and to implement those decisions at the national level." Such a cohesive strategy, however, does not appear to be forthcoming.

In Geneva, Undersecretary Wirth's announcement that the United States would require mandatory oil- and coal-burning reductions became the basis of a ministerial agreement that was hammered out at a late-night meeting of delegates from the United States, the European Union, Brazil, and China. Although the goals it set were modest, it did specify legally binding emissions limits. One official told a Reuters reporter that should the agreement be approved by the conference, the conference delegates would leave Geneva with instructions to "tell their governments to accelerate negotiations for a legally binding target."

Despite its mildness, the ministerial agreement provoked an escalation of hostilities from the fossil fuel producers. It was voted down in Geneva, opposed by fourteen oil- and coal-producing nations, including most of OPEC, Russia, and Australia. "I'm disappointed that the Americans have insisted on binding limits," said Australian prime minister John Howard. "It hurts Australia for the Americans and the Europeans to take the line they have taken, and I'm disappointed it's occurred." Then he passed the buck to the developing world: "You won't get serious worldwide action on this unless you get some constraints operating on greenhouse gas emissions in the developing countries."

Two weeks later, the oil lobby took a similar line in a Mobil ad that appeared on the op-ed page of *The New York Times.* Legally binding emissions targets, it declared, are "likely to cause severe economic dislocations. . . . Though the industrialized world accounts for half of greenhouse gases, that share will drop as developing nations flex their economic muscle. If developed nations act *alone* to reduce emissions, the staggering cost . . . will drive nations to export much of their industrial base to countries with less stringent controls. . . . The dislocations will be even more severe if the solutions are not implemented globally."

It's the same old strategy—the one Kevin Fay called a nonstarter: If you saddle the developing nations with the same emissions-reduction burden as the wealthier ones, you will scuttle any meaningful participation by two-thirds of the world.

Virtually no one outside the fossil fuel community believes there is any real solution to the global climate threat short of a change in the world's basic energy diet. Given the growing momentum toward development in the poor nations, that change must be a worldwide one if it is to avert crippling climate disruptions.

The obstructionism of the oil interests aside, what has been conspicuously lacking in the climate negotiations is leadership on the part of the United States. Fearing the country's ascendant conservative political mood, epitomized by anti-UN paranoia, U.S. officials have been reluctant to address the underlying economic inequity between the rich and poor countries, as well as their massive inequity in energy resources. The Clinton administration has failed to explain to the electorate why American economic survival depends on making a substantial correction in this inequity.

As the economic journalist Robert Kuttner told the United Nations last year, "Both rich and poor nations have a common stake in policies that put the globe on a sustainable development path. The conflict is less between poor and rich countries than between the broad interests of people and the narrow interests of extractive industries. We need to find our way towards some kind of global regime that reduces emissions of greenhouse gases, but well-off nations need to transfer the technology to make this possible, rather than viewing this shift as one more opportunity for private industry to profit."

But America's leaders must make the case, in much the same way that President Truman made the case for the Marshall Plan to rebuild Europe after World War II. Truman's central selling point was that the Marshall Plan was critical to the United States' own national security.

"We need to explain why a transfer of wealth to the developing world—in the form of alternative energy technologies—is in our country's own basic self-interest," says William Ruckelshaus. "That is the key. We need to think of sustainable development as something that is not only compatible with the needs of the planet. We also need to see that it is consistent with our own sense of economic and political freedom. We should promote economic growth—but only in the context of the underlying environmental goals our people stand for. Environmental protections are embedded in our national values. To fund the transfer of sound energy technologies to the developing world is not some fuzzy-headed, liberal, do-gooder notion. It is critical to our own economic and environmental self-interest.

"In the absence of such a transfer, if the developing countries take the same [fossil fuel] path of development we did, it will be to the detriment of our own country and our own security. That's the way to think about it," he adds.

To the Philippines' LaVina, the issue transcends national self-interest. "Ultimately, the question goes beyond even science and technology," he says. "Ultimately the question is one of ethics. The emergence of climate change poses a question to all of us about what kind of world we want to live in. If I had a magic wand and could transform the way we live, I would decree that all of humanity become far less wasteful, far less greedy, far less avaricious. To be frank, this is not just a North-South issue. The North has no monopoly on wastefulness and greed. It is a human problem. Ultimately, it is a question of ethics."

Argentine climate negotiator Raul Estrada Ayala puts it differently. Referring to all the nations of the world, he says, "We are all adrift in the same boat. And there's no way that only half the boat is going to sink."

On April 22, 1996, more than 100 people were killed in heavy flooding across Afghanistan. At least seven provinces, covering more than a quarter of the country, were affected by the floods, brought on by melting snow and heavy rains. In northern Badakhshan, the villages of Joma Bazar and Islim were virtually washed away when a giant wall of water came crashing down on the area, one relief worker said. At least 15 people are known to have died and 25 others are missing. Hundreds of homes have been destroyed and hundreds of cattle killed, he said. Each year the arrival of spring melts the snow on the vast Hindukush mountain range, causing rivers to swell and burst their banks. But in 1996, the heavy rains combined to produce the worst flooding in decades.

Three months later, 17 inches of rain fell in 24 hours in Aurora, Illinois, while 10 inches fell in the nearby city of Naperville, causing extensive flood damage to nearly 8,000 homes. Governor Jim Edgar declared 13 counties state disaster areas and mobilized the National Guard to help in the massive cleanup. The flooding closed roads and businesses and, when a train line had to be shut down, stranded tens of thousands of commuters. "I've never seen it rain like that for so long, and I've been in business here for thirty years," said Don Feldcott, owner of the Lantern restaurant and bar in Naperville. "We had six feet of water in the basement. Everything is ruined. The water turned two five-foot refrigerators upside down." Naperville officials put the damage at $30 million and counting.

The following month, 62 people died and more than 180 were injured when a torrent of mud and rock swept over a crowded family campsite in the Pyrenees mountains of northern Spain. Bodies were pulled from the mud as far as 10 miles downstream from the campsite, officials said. "I have flown over the area—it's a dreadful sight," said Prime Minister José Maria Aznar, who interrupted a beach vacation to survey the damage. Rescue teams said they feared that dozens more bodies could still lie downstream in a river near the Virgen de las Nieves (Virgin of the Snows) campsite in Biescas, a mountain town about 80 miles east of Pamplona. "It all happened in a flash—I can't explain it. It was like a giant wave carrying off everything—the cars, the trailers," one survivor told Spanish television. "It was a matter of seconds, not even minutes. The main street in the campsite was a river of mud, between three and six feet deep." Wrecked cars, flattened caravans, and other debris from the site, which was at full capacity with around 700 tourists, were carried more than half a mile away.

SIX

Headlines from the Planet

*OUR ELECTED REPRESENTATIVES IN WASHINGTON CON-*tinue their assault on knowledge by cutting away at the foundations of scientific research. The world's diplomats promote their delicately balanced and multilayered agendas at negotiating sessions. The captains of industry jockey for ways to suppress news of climate change or turn the issue to their own advantage. Environmental groups compete with their corporate and ideological adversaries for the attention of an apathetic public. But soon, it appears, the immense changes taking place in the global climate will drown out all these human machinations.

Far from the computer-modeling laboratories of the IPCC scientists and independent of freakish weather extremes, a series of deeply disturbing physical changes in the oceans, forests, glaciers, and plains are signaling the increased stresses on the earth.

Pretend, for a moment, that you are the editor of a periodical called *Planet News.* Your correspondents report from outer space (with its clarified perspective of our suspended globe), from the earth's high mountains, from its ancient forests, and its warming oceans. In putting together the next issue, you read over some of the items that will likely warrant headlines:

OCEAN WARMING CREATES PACIFIC WASTELAND

Three months after the fracturing of Antarctica's Larsen ice shelf, researchers announced that the population of tiny marine organisms called zooplankton off the coast of southern California had declined by a stunning 70 percent over the last twenty years. Scientists linked the decline in zooplankton, a food source for several species of fish in the region, to a 2-to-3-degree-Fahrenheit increase in the area's surface water temperature over the last four decades. The warming seas have created a vast wasteland, with few birds and few fish. John McGowan, a biologist at the Scripps Institution of Oceanography and co-author of the March 1995 zooplankton study, recalled observing abundant fish and bird life during his cruises in the area in the 1960s. But, he told the Associated Press, during a recent voyage he was "flabbergasted to see the difference." McGowan said the startling decline in the zooplankton population raises questions about the survival of sardines, anchovies, hake, and Pacific mackerel. The rising water temperature, he said, was robbing surface waters of nutrients like nitrates and phosphates that the plant plankton need to survive. Because zooplankton feed on plant plankton, the loss of nutrients ripples up the food chain. In a March 1995 article in the journal *Science*, McGowan wrote that over the past few decades a significant warming in the top six hundred feet of water has led, among other things, to the collapse of commercial anchovy fishing in the area. Dr. Dick Veit, a zoologist at the University of Washington, said McGowan's findings were consistent with other studies that had shown stunning losses of fish and seabird populations along the Pacific Coast. Veit said he had discovered a 90 percent decline in the population of one group of seabirds in the region.

SMALL TEMPERATURE RISE FUELS MIGRATIONS OF SEA ANIMALS

Around the same time that scientists announced the decline in the Pacific zooplankton population, other researchers noted that rising water temperatures in Monterey Bay have triggered an exodus of cold water crabs, snails, and other species and an influx of populations of different sea animals accustomed to warmer temperatures. Stanford University biologist Chuck H. Baxter said the findings indicate "there has certainly been a climate change at least at that site." The changing population of snails, crabs, and other creatures was discovered when researchers repeated a survey of forty-five species of sea animals that was first conducted sixty years earlier, in a part of a cove near the Hopkins Marine Station in Monterey Bay. Baxter, co-author of a study published in February 1995 in the journal *Science,* said the new survey, conducted in 1993 and 1994, found that northern species were in decline and the abundance of southern species had increased.

"The most dramatic change was in a snail that was not even detected in the earlier study," said Baxter. The snail is found on rocks in the intertidal areas of the cove. Another species, a predatory snail, had been rare in the 1930s survey but now was abundant, he said. Species of crab and starfish that prefer cold water were severely reduced in number. A southern sea anemone had moved in vigorously, while a northern variety was in decline. Also, a type of algae that prefers cold water had declined, while a southern species of the plant bloomed. Baxter said instruments at the site indicate that the annual mean temperature on the shoreline at the cove has increased by about 1.3 degrees Fahrenheit over sixty years and that the maximum summer temperatures have increased by about four degrees. The temperature differences are small, Baxter said, but they are enough to fundamentally change the animal and plant communities in the cove.

"These species changes . . . show that even a modest change in temperature can cause organisms to respond," he said. Subtle changes may be under way worldwide. Research should be conducted wherever a record exists for a comparison of species types and numbers, Baxter said.

His observations echo those of Harvard University's E. O. Wilson and other prominent biologists concerned about the devastation of diverse life-forms by climatic changes. Wilson has documented the fact that the number of species becoming extinct is accelerating at an unprecedented rate. And Dr. Thomas E. Lovejoy, of the Smithsonian Institution, wrote recently: "Among the repercussions the greenhouse effect is likely to have, the hardest to mitigate is the loss of biological diversity. . . . Today biological diversity is signaling that the sheer numbers of people combined with our effects on the environment have almost reached the point of no return."

BUTTERFLY STUDY CONFIRMS WARMING-DRIVEN MIGRATIONS

While previous studies had documented only localized species migrations in response to atmospheric warming, an August 1996 study published in the journal *Nature* indicated that the entire migratory range of one type of butterfly has shifted northward in response to just a slight increase in temperature. Following a fourteen-month survey of Edith's checkerspot butterfly populations, Dr. Camille Parmesan discovered that butterfly colonies were becoming extinct at the southern end of their geographic range in Mexico and southern California, while their numbers were increasingly significantly in Canada. "What my study shows," she told *The New York Times*, "is that we're apparently seeing an effect, even though this is a small warming. The effect is stronger than I expected." "When you put it together," she added, "the picture that's coming out is that climatic warming is affecting wild organisms." Parmesan's butterfly study is the first that surveyed the entire geographical range of a species' migration pattern. While other studies documented local population shifts, such as the northward migration of warm-water sea animals in the Pacific Ocean, they were too localized to show a range shift for the species as a whole. According to Parmesan, a researcher at the National Center for Ecological Analysis and Synthesis at the University of California at Santa Barbara, the southernmost butterfly colonies were four times more likely to be extinct than those at the northernmost latitudes.

MELTING OF THE WORLD'S GLACIERS ACCELERATES

The vast majority of the earth's glaciers have been melting for at least two decades. The most recent measurements indicate that the rate of so-called "glacial retreat" is accelerating rapidly. "Between one-third and one-half of existing mountain glacier mass could disappear over the next hundred years," the IPCC declared in 1995.

One indication is that the volume of the world's half-million small glaciers, such as those found in the Alps, has shrunk markedly in the last century—and some scientists suspect that water from the melting mountain ice has raised sea levels by nearly half an inch since 1960. Glaciologist Mark Meier, of the University of Colorado, noted that small glaciers have lost more than 10 percent of their total mass in this century—while alpine glaciers have lost almost 50 percent of their ice due to rising global temperatures. Presenting his findings at a 1995 conference of the International Union of Geodesy and Geophysics, Meier theorized that water loss from small glaciers has accounted for a significant portion of the observed sea level rise over the last twenty years.

Halfway around the world, Australian researchers discovered recently that glaciers in Indonesia have receded at a rate of 45 meters a year over the past twenty years. Comparing current measurements to photos and maps from the 1930s, researchers James Peterson of Monash University in Melbourne and Geoffrey Hope of the Australian National University calculated that the rate at which the midlatitude glaciers are retreating accelerated from 30 to 45 meters a year between 1971 and 1993.

Nor is the glacial retreat confined to warmer regions. Strong indications of warming have been found in ice core drilling in glaciers on the Tibetan plateau, which indicate that the last fifty years appear to be the warmest in the last 12,000 years. That assessment comes from two of the world's leading glaciologists, Drs. Lonnie G. Thompson and Ellen Mosley-Thompson at Ohio State University. The greatest rate of retreat they discovered is occurring in the largest glacier in the Peruvian Andes. The speed with which it retreated between 1963 and 1983 has tripled between 1983 and 1991, with a sevenfold increase in the rate of loss of its volume of ice.

Between 1963 and 1987, moreover, the total ice cover on Mount Kenya in Africa decreased by 40 percent.

Noting the local impacts of the glacial retreat, the authors wrote: "The loss of these valuable hydrological stores may result in major economic and social disruptions in those areas dependent upon the glaciers for hydrologic power and fresh water."

The glacial ice cores, moreover, provide an invaluable record of past temperature changes in the earth's climate. But Thompson and Mosley-Thompson believe that that record is being obliterated by the warming of the atmosphere. In a 1995 report they explained that ice cores "from diverse locations, such as Antarctica, Greenland, the Tibetan Plateau and the Andes . . . have been [used] to produce a high resolution global perspective of climate for the past 1000 years. . . . Unfortunately, due to the unprecedented warming occurring at high elevations in the tropics and subtropics, many of these valuable archives are in imminent danger of being lost."

The retreat of tropical glaciers correlates closely with rising tropical sea surface temperatures, according to findings by Henry F. Diaz of NOAA and Nicholas E. Graham of the Scripps Institution of Oceanography. Noting recent changes in the heights of freezing-level temperatures, the authors wrote in *Nature* that "the warmth recorded in the tropical oceans in several past decades may perhaps be at an unprecedented level since the mid-Holocene period. . . . Recent evidence . . . suggests that high-elevation environments may be particularly sensitive to long-term changes in tropical sea surface temperatures and tropospheric humidity, which are likely to have an impact on the hydrological and ecological balances of high-altitude zones throughout the globe perhaps to the greatest extent in the tropics."

FOREST GROWTH STUNTED BY INCREASED HEAT, CO_2

Researchers studying the effects of warming on the northern Canadian and Alaskan forests—where temperatures have warmed by 2 degrees Celsius over the last 90 years—have found that after an initial spurt, the trees' growth rate flattened. The type of forest they studied—boreal—is the earth's largest terrestrial ecosystem and constitutes a third of all the world's forests. The scientists had expected that a warmer climate, with its diet of enriched CO_2, would increase the trees' growth rates and expand the northern edge of the Canadian forests farther into the Arctic. But researchers Gordon Jacoby and Roseanne D'Arrigo of the Lamont-Doherty Earth Observatory, who studied growth rings in trees near the timberline in northern and central Alaska, reported in May 1995 that while the growth rate accelerated during the 1930s and 1940s with warming weather, it has subsequently plateaued. "The warming appears to be stressing the forests by speeding moisture loss and subjecting them to more frequent insect attacks," said Jacoby. If there's enough moisture, he added, the warmer temperatures help these trees to grow better. But if it's too warm, the evaporation of tree water through its leaves and needles reverses that benign effect. Warm sunny days draw the trees' moisture out through their leaves and needles. If there is insufficient rain or soil moisture to compensate, tree growth will slow. The scientists' strongest worry, however, is that climate warming in Alaska will spur a dramatic increase in the insect population. "Alaskan forests have suffered from severe outbreaks of bark beetles, which have devastated several million acres of forest. Entomologists have pointed out that warmer temperatures can halve the reproductive cycle of the bark beetle from two years to one," according to the researcher.

ARCTIC WARMING REVEALED IN SOIL, SURFACE, AND OCEAN MEASUREMENTS

North of the Canadian forests, a series of boreholes in Alaska revealed soil temperature increases of 2 to 5 degrees Celsius (3.6 to 9 degrees Fahrenheit) during this century. Researcher David Deming noted that while such increases in underground temperature are not proof of atmospheric warming, it does appear the most likely of all possible causes. "The solid Earth . . . continuously filters out daily and seasonal changes in ground surface temperatures while maintaining a running record of the longest mean. . . . The sum of evidence is consistent with theoretical predictions of warming related to the accumulation of greenhouse gases in the Earth's atmosphere from [human] activities." Deming's 1995 findings parallel a discovery the previous December that a deep layer of the Arctic Ocean has warmed by 1 degree Celsius in the last few years. Around the same time, scientists at the National Oceanic and Atmospheric Administration discovered that surface temperatures at nine stations north of the Arctic Circle have increased by about 5.5 degrees Celsius (9.9 degrees Fahrenheit) since 1968.

HEAT-ENHANCING VAPOR INCREASES IN UPPER ATMOSPHERE

In one of the more ominous recent atmospheric findings, NOAA researchers announced in March 1995 that they had found evidence of enhanced atmospheric heating in the form of "significantly increased" amounts of water vapor in the lower stratosphere, above the level at which sulfur aerosols mask the warming with their heat-reflecting action, just beyond the twelve-mile limit of our atmosphere. "The detection of an increase of water vapor in the stratosphere [between 1981 and 1994] adds to the concerns about the build-up of greenhouse gases and their effects on atmospheric chemistry as well as climate," the scientists wrote. Water vapor is generated, at lower levels, by atmospheric warming. At higher levels it enhances the heating process. While the direct measurements, conducted by balloon-borne instruments, were taken over Boulder, Colorado, they "are expected to be representative of the stratosphere over the . . . northern midlatitudes." The increase in stratospheric water vapor—which intensifies lower-level warming—over the belt of earth that is home to most of the industrialized countries was, they said, "larger than might be expected and may be linked to observed global temperature rise in recent decades."

RECENT EL NIÑO A ONE-IN-TWO-THOUSAND-YEAR EVENT

Much of the warming of the seas—especially in the Pacific Ocean—is attributable to a phenomenon called El Niño, a pool of warm water that periodically surfaces in the tropical latitudes of the eastern Pacific. The effects of El Niño are far-reaching. They fuel violent storms in the Pacific, floods across California and the U.S. Gulf Coast, and droughts in Australia and Africa. But while most El Niño events last only a year or two, scientists noted at the beginning of 1996 that the recent El Niño—which had lasted for five years and eight months—was a one-in-two-thousand-year occurrence. In January 1996 researchers Kevin Trenberth and Timothy Hoar, of the National Center for Atmospheric Research, noted that the event—coupled with the past two decades of sea surface temperature anomalies—is "unexpected given the previous record, with a probability of occurrence about once in 2,000 years. This opens up the possibility that the . . . changes may be caused by the observed increases of greenhouse gases."

Historically, El Niños have alternated every two or three years with La Niñas (upwelling pools of cool water in the eastern tropical Pacific), in what scientists call the Southern Oscillation. That oscillation is part of a large, complex interaction between the Pacific Ocean and the global atmosphere. But over the past twenty years, there has been only one significant La Niña. By contrast, the strongest El Niño of the century occurred in 1982–83, and the longest-lived one concluded in 1995. Trenberth and Hoar pointed out that the latest El Niño "occurred in the context of a tendency for more frequent El Niño events and fewer La Niña events since the late 1970s. . . . These results raise questions about the role of climate change. Is this pattern of change a manifestation of the global warming and related climate change associated with increases in greenhouse gases in the atmosphere? Or is this pattern a natural decadal-timescale variation? We have shown that the latter is highly unlikely."

NORTH ATLANTIC OCEAN WAVES 50 PERCENT HIGHER

While some scientists say it is too early to prove definitely that atmospheric warming is contributing to an increased intensity of storms, researchers at the Institute of Oceanographic Sciences in Britain discovered that average waves in the North Atlantic around the northern tip of Scotland were 50 percent higher in 1993 than they were during the 1960s, with storm waves averaging 10 percent higher. "We have had a whole series of very deep depressions crossing the Atlantic this winter," said Barry Parker of the British Meteorological Office. "It all seems to be tied up with a steepening gradient in sea temperatures." He was referring to the difference in temperatures between the icy waters between Iceland and Greenland and the warming seas farther south, near the British Isles. Researcher John Gould noted in 1993 that "there is evidence we seem to be entering an era of strong storms which might be associated with global warming."

U.S. WHEATFIELDS COULD BE DESERTS IN A DECADE

The great crop-growing plains in the midwestern United States are far more susceptible to desertification from temperature change than was previously believed, according to 1996 findings by researchers at the U.S. Geological Survey. The team of scientists, led by Dr. Daniel Muhs, have discovered that the western plains, covered by a thin layer of grazing grass, could become deserts after a few years of drought conditions. The findings, which indicate that our current Holocene climatic period has been much more volatile than previously believed, imply that relatively small changes in atmospheric circulation patterns—such as those projected under increased greenhouse conditions—have triggered massive floods as well as the migration of sand dunes over previously fertile growing land. The catastrophic Sahara-like conditions that occurred in the western plains as late as the eighteenth century "could be touched off if predicted climate changes resulting from accumulations of heat-trapping gases like carbon dioxide come to pass," according to reporter William K. Stevens of *The New York Times.*

DESERT CONDITIONS SPREADING IN SOUTHERN EUROPE

The process of desertification has already been under way for nearly thirty years in parts of Spain, Portugal, Greece, and Italy, according to a 1996 report by more than forty European climate scientists working under the auspices of the European Commission. The scientists found during their five-year study that protracted droughts, punctuated by intense, soil-eroding downpours, are becoming the norm rather than the exception, as the earth's atmosphere continues to warm. John Thornes of Kings College, London, who coordinated the study, told *New Scientist* magazine that scientists discovered that the extended drought in Spain, between 1990 and 1995, was only part of the trend. The group's new analyses of rainfall data revealed "a turning point toward progressively lower rainfall since 1963," he said. At the same time, he added, there has been "a clear increase in the number and duration of both heat waves and violent storms." The announcement of the scientists' findings followed by two months a declaration by the UN Food and Agriculture Organization (FAO) that "the sustainability of Mediterranean agriculture appears questionable unless urgent and drastic measures are taken."

RISING TEMPERATURES BRING EARLIER SPRINGTIMES

Spring is now arriving a week earlier in the Northern Hemisphere than it did twenty years ago, and rising atmospheric temperatures are the most likely cause, according to researchers at the Scripps Institution of Oceanography. Their findings indicate that climate change is measurably altering the earth's carbon dioxide cycle. Charles Keeling, an author of the study, noted that "our measurements of the increase in seasonal changes of carbon dioxide levels conform very well with temperature measurements going back to the late 1950s showing that surface temperatures are getting warmer." In addition, Keeling noted, "Small changes in temperature—averaging only a fraction of a degree Celsius a year—are nevertheless causing fairly large changes in plant growth."

Each spring and summer the level of carbon dioxide in the atmosphere drops as plants absorb the gas, then rise again as plants die and decay, releasing carbon dioxide during fall and winter. According to the study, the Scripps researchers found that since the early 1960s, seasonal swings in carbon dioxide concentrations have grown 20 percent larger in Hawaii and 40 percent larger in Alaska.

One result is that the spring decline in carbon dioxide now begins seven days earlier than it did during the mid-1970s. "The drawdown in carbon dioxide is earlier than it was before, and that's probably the key to the whole picture, because it looks like the growing season has lengthened," Keeling told *Science News.*

SCIENTISTS DISCOVER FURTHER DISINTEGRATION OF ANTARCTIC SHELVES

In the year following the breaking-off of a giant piece of the Larsen ice shelf, scientists discovered that five of the nine ice shelves attached to the Antarctic Peninsula have disintegrated over the last half century, as temperatures have risen.

Researchers David Vaughan and C. S. M. Doake, of the British Antarctic Survey, wrote in January 1996 that surface temperatures in the Antarctic Peninsula have increased by 4 to 5 degrees Fahrenheit in the last fifty years. That temperature increase—which is five times more than the worldwide increase—has caused five coastal ice shelves to break up.

"People talk a lot about global warming in terms of average temperature," Vaughan told *The Boston Globe.* "But the impact is going to be local areas where the climate changes very significantly. If Antarctica experiences above-average temperature increases," he added, "it could begin a process that significantly raises the earth's sea level." If the polar ice sheet continues to break up, it would add significantly to the three-foot sea level rise already predicted by the world's climate scientists for the next century.

Vaughan and Doake caution that if the warming continues, it could trigger the melting of ancient ice sheets in the interior of Antarctica.

H. Jay Zwally, a glaciologist at the Goddard Space Flight Center, told the *Globe,* "If this is a precursor of what might happen on an expanded scale in Antarctica, it would become far more serious in the future." He pointed out that at the end of the last ice age ten thousand years ago, melting ice raised sea levels by thirty feet in only a few hundred years.

While scientists do not agree on the rate and extent of global warming caused by burning fossil fuels, Dr. Rodolfo del Valle, of Argentina's National Antarctic Institute, thinks greenhouse gases may be dangerously accelerating a natural warming process in the region. "I think the process is like a father pushing his child in a swing. If the father pushes too hard, the swing twists around, the child falls down, and the game is over."

CLIMATE CHANGES FUEL SPREAD OF OLD, NEW DISEASES

While the changes noted in these articles affect life in the oceans, the mountains, and the forests, another set of findings over the past three years extends those changes to the realm of human health. In December 1995 the IPCC working group on climate impacts declared, "Climate change is likely to have wide ranging and mostly adverse impacts on human health with significant loss of life."

It already has had such impacts. Among the various effects of climate change, extended heat waves pose the most obvious threat to health—as in the hundreds of heat-related deaths in Chicago and India during the summer of 1995. But a far more insidious threat is the spread of infectious diseases, most notably malaria, dengue, yellow fever, cholera, hantavirus, and encephalitis.

In January 1996 a report in the *Journal of the American Medical Association* indicated that there is a correlation between climate change and the spread of infectious disease. It followed a series of similar reports in 1994 in the *Lancet,* a prestigious British journal of medical research. And in July 1996 a panel of the World Health Organization, the World Meteorological Organization, and the UNEP released a report warning that the impacts of climate change so threaten human health that "we do not have the usual option of seeking definitive empirical evidence before acting. A wait-and-see approach would be imprudent at best and nonsensical at worst."

As we have seen, the *Aedes aegypti* mosquito, which carries dengue and yellow fever, has traditionally been unable to survive at altitudes above 1,000 meters because of the colder temperatures there. Recently, however, this mosquito has been reported at 1,240 meters in Costa Rica and as high as 2,200 meters in Colombia, according to Dr. Paul Epstein of the Harvard Medical School and a contributor to the IPCC. It is also appearing at higher elevations in Uganda, Rwanda, Kenya, and Ethiopia, where residents recently suffered outbreaks of yellow fever. The ascent of the insects' habitats matches the upward migration of mountain vegetation. Malaria has also established itself in high-altitude zones of Rwanda, where it was previously unseen. Several years ago, during a heat wave in Mexico, outbreaks of dengue fever occurred for the first time at altitudes as high as 1,700 meters.

In 1995 researchers in the Netherlands and Great Britain estimated that if the IPPC's projected level of warming is realized, the "epidemic potential of the mosquito population" in the tropical regions would double. In the temperate climates—home to the United States, most of Europe, the former USSR, China, and Japan—it would increase a hundredfold. "There is a real risk of reintroducing malaria into nonmalarial areas, including portions of Australia, the U.S., and southern Europe," the researchers concluded. An increase of 3 degrees Celsius could cause an additional 50 to 80 million cases of malaria each year around the world. The situation would be especially grim in areas where people have acquired little or no immunity to the disease.

The spread of dengue and yellow fever parallels the spread of malaria. Unusual rain patterns triggered outbreaks of dengue recently in Costa Rica and Brazil. In 1995 a more virulent form of the disease—dengue hemorrhagic fever—was discovered to be moving north, raising concerns about outbreaks in the United States. In its less severe form dengue, which is also known as "breakbone" fever, is marked by high fevers, extreme chills, diarrhea, and vomiting, followed by severe headaches and extreme bone and joint pain. When it abates after several weeks, it normally leaves the depleted victim in a state of depression. In its more virulent hemorrhagic form, it produces bleeding that first appears as red spots on the legs and, in lethal cases, leads to hemorrhaging of the lungs, brain, and gastrointestinal system. A recent epidemic in Puerto Rico infected 15,000, raising fears about outbreaks in Florida, Texas, and other areas in the southern United States. Most alarmingly, the average incidence of hemorrhagic dengue—just over 100,000 cases a year worldwide between 1981 and 1985—quadrupled between 1986 and 1990, averaging to more that 450,000 cases a year.

Changing climate conditions have also been linked to the spread of cholera. Following an outbreak in January 1991 near Lima, Peru, the disease spread north to the Peruvian port of Chimbote. Over the next week 12,000 cases were recorded. The epidemic then spread along the Peruvian coast, breaking out in three separate ports within three weeks. By the fourth week, it was being transmitted by travelers into the interior mountain regions. Continuing its waterborne migration, it spread into Ecuador at the end of February and into Colombia eight days later. In mid-April health officials in Chile reported outbreaks of cholera, and on April 22 it appeared in western Brazil, in a region that shares the Amazon headwaters with Peru. All together, in the eighteen months following the initial outbreak, more than a half million people had become infected and more than 5,000 had died in Peru, Ecuador, Colombia, Guatemala, Mexico, Chile, Panama, and Brazil.

Dr. Rita Colwell, director of the University of Maryland's Institute of Biotechnology, attributes the spread of the disease to two factors. One is the explosive growth of coastal algae blooms spurred by warming surface waters and upwellings of nutrients released by shifts in ocean currents. The second is unseasonal and severe flooding, which spreads sewage into local water supplies. Both are consequences of climatic instability. In the two years following the Latin

American epidemic, cholera outbreaks appeared in Africa, India, and Bangladesh.

Nor are the disease-spreading effects of climate change confined to outbreaks of bacterial diseases in developing countries. In 1993 a rodent-borne virus—hantavirus—broke out in the southwestern United States following a period of extreme changes in the local climate. For six years the area had experienced a prolonged drought, during which the lack of water virtually eliminated the population of animals that prey on desert rats and mice—snakes, owls, and coyotes. Then an unusually prolonged period of heavy rains in mid-1992, led to an explosion of pinion nuts and grasshoppers—both of which are nutritious rodent foods. In only one year the southwestern rodent population increased tenfold, in the process infecting the human inhabitants of the area. "Hantavirus was amplified and transmitted to humans as a consequence of environmental changes driven by a change in climate and climate variability," Paul Epstein explains, adding that climatic extremes are also associated with such diseases as shigellosis and hepatitis in South America, viral encephalitis in Australia, and eastern equine encephalitis in Massachusetts. Eastern equine encephalitis is relatively rare in humans, but it kills about 60 percent of those who contract the disease, while the remainder suffer permanent neurological damage. Dr. Jerome Freier, of the Department of Agriculture's Center for Epidemiology and Animal Health, notes that most cases occur near wooded areas adjacent to swamps and marshes—the disease is borne by mosquitoes. Rising temperatures, combined with heavy rainfall, could extend the active season of the mosquitoes and provide them with larger areas of swampland in which to propagate. The disease is most pervasive in the Southeast, as well as in Massachusetts, New Jersey, and New York.

Do we need more evidence that climate change is affecting the planet? The human population has become so huge and our technology so powerful that the weight of our industrial emissions—amplified by the cutting down of CO_2-absorbing trees and forests—is having impacts on the earth in ways we never intended. We are twisting our planet out of shape. We didn't mean to, but we are.

As the nations of the world convene in Kyoto, Japan in late 1997 to negotiate an international climate treaty—with emissions reduction levels and timetables—they will also be calculating the costs of such reductions to their own industries, the possible shifts in their economic relationships with other countries, and the requirements of their domestic political constituencies. Meanwhile Antarctica is fracturing. The ocean off California is becoming a wasteland. Plants are migrating up mountains to keep pace with the warming climate. Whole species are migrating under the surface waters of the seas. The Arctic soil under the north pole is warming. The oceans are rising, and their waves are taller. High in the mountains, the glaciers are shriveling. Forests are rapidly losing their ability to thrive. And insects are poised for a population explosion that would jeopardize our food crops, our trees, and our health.

It is this knowledge that should dictate the purpose of the delegates to the Climate Convention. If it does not—if they succumb to the fossil fuel lobby or the short-term political considerations of their own governments—if, in short, they treat the negotiations as yet another occasion for diplomatic jockeying—the changes in the planet will only escalate, and the political world will undergo transformation as well.

In Yemen in June 1996, more than 330 people died during the worst floods in nearly 40 years. The floods left thousands of Yemenis homeless and caused more than a billion dollars in damage. The following month, pools of stagnant water that had been left in the wake of the floods gave rise to an explosion in the mosquito population. In the Yemeni governate of Marib, 30 people died from the resulting malaria epidemic. Half the area's population—about 168,000 people—were infected with the disease.

SEVEN

The Coming Permanent State of Emergency

LONG BEFORE THE SYSTEMS OF THE PLANET BUCKLE, DEmocracy will disintegrate under the stress of ecological disasters and their social consequences.

Two different men independently expressed this chilling insight to me—William Ruckelshaus, the first head of the EPA and now CEO of Browning-Ferris Industries; and Dr. Henry Kendall of MIT, the recipient of the 1990 Nobel Prize for physics, who has devoted his retirement to the study of critical global trends.

When I first heard the remark, it seemed shocking yet somehow irrelevant to the climate crisis. Only after the thought had burrowed its way into my consciousness did the connection became apparent: If we alter the balance of natural relationships that support our lives, those changes will ripple through the complex relationships that make up our society.

The question that struck me next was why this understanding has played no part in the climate debate. Environmentalists naturally focus on the ecological consequences of climate change—not on its impacts on civilization. Scientists, struggling with the urgent demands of their mission, focus on remedying the gaps in their knowledge and improving their methodologies—not on the political and economic consequences of their discoveries. And business leaders, insofar as they think about climate change at all, focus on its potential impact on their competitive position or its ability to affect their market share—not on the delicate balance between a democratic society and a corporate state.

Government leaders rarely address problems that cannot be solved before their next reelection campaign. The shelf life of many political issues is four years. In this age of instant response, businesses likewise are largely captive to the short-term demands of their shareholders and directors. In our late twentieth-century fast-forward world, institutions respond to events only when they reach emergency proportions. Yet given the long atmospheric lifetime of greenhouse gases, the destructive instabilities of the global climate will continue long after we reduce our emissions—as we finally must.

In the United States the mere threat of impending climate change has impelled the oil and coal industries to engineer a policy of denial. While their campaign may seem at this point no more sinister than any other public relations program, it possesses a subtle antidemocratic, even totalitarian potential insofar as it curbs the free flow of information, dominates the deliberations of Congress, and obstructs all meaningful international attempts to address the gathering crisis. The stress caused by climate change is lethal to democratic political processes and individual freedoms.

For MIT's Kendall, it is the poor, precarious, nations of the developing world that would face the threat of totalitarianism first. In many of these countries, where democratic traditions are as fragile as the ecosystem, a reversion to dictatorship will require only a few ecological states of emergency. Their governments will quickly find democracy to be too cumbersome for responding to disruptions in food supplies, water sources, and human health—as well as to a floodtide of environmental refugees from homelands that have become incapable of feeding and supporting them.

Diminishing food and water supplies already pose a grave threat to the survival of democracy in the developing world. As climate instability intensifies, that threat is bound to become reality. "The world's food supply," says Kendall, "must double within the next thirty years to feed the population, which will double within the next sixty years. Otherwise, before the middle of the next century—as many countries in the developing world run out of enough water to irrigate their crops—population will outrun its food supply, and you will see chaos. All we need is another hit from climate change—a series of droughts or crop-destroying rains—and we're looking down the mouth of a cannon."

But even before the ravages of climate change have become widespread, Kendall believes, it may already be too late to head off pervasive famine and social disruption. Until recently, global food production has kept up with population (except in Africa), in large part because of the Green Revolution. The Green Revolution is a program introduced in the 1960s to enhance crop-seed production. It requires intensive use of fossil energy for fertilizers, pesticides, and irrigation. Its conceiver, Norman Borlaug, intended the Green Revolution to be a short-term effort by which poorer countries could develop modern, sustainable growing practices. Unfortunately, as Kendall and David Pimentel, a professor of agriculture sciences at Cornell University, point out, "the Green Revolution has been implemented in a manner that has not proved environmentally sustainable. The technology has enhanced soil erosion, polluted groundwater and surface water resources, and increased pesticide use has caused serious public health and environmental problems."

Such erosion is causing a growing scarcity of arable land in many parts of the world. It takes, on average, five hundred years for the earth to form one inch of new topsoil. Today, erosion is removing between 10 and 120 tons per acre of topsoil each year around the world. This accelerating loss of nutrient-rich soil is far outpacing the rate at which it can be replaced. In addition to erosion, much arable land is being lost to salinization from rising salt tables and waterlogging from improper irrigation.

The outcome of this deterioration is loaded with totalitarian potential. In the mid-1980s a thirty-year growth in global food supplies reached its peak. Food production is now declining. Today only two of the world's 183 nations—the United States and Canada—are major exporters of grain. Yet as the world's population expands at an almost exponential rate, the earth is losing nearly 1 percent of its agricultural lands every year.

Nor is the situation with available water much better. In many arid parts of the world, freshwater resources are becoming overtaxed. They are depleted by industrial overuse and by the demands of growing concentrations of people in the cities of the developing world. United Nations secretary-general Boutros Boutros-Ghali was deadly serious when he noted, a few years ago, that the next war in the Middle East will be fought not over oil but over water.

To Americans, an adequate food supply seems a basic human right. To many non-Americans, climate change threatens to turn it into an object of fantasy. Malnutrition, illness, starvation, and crime are all socially destabilizing consequences of food shortages. Kendall and Pimentel warn that demands on the food supply will be immense in the coming years. The next century's expected "doubling of the population would necessitate the equivalent of a tripling or more of our current food supply to ensure that the undernourished were no longer at risk and to bring population growth stabilization within reach in humane ways—without widespread hunger and deprivation."

The catalog of food and water deficits cited by Kendall, Pimentel, and other food-supply specialists is staggering. But the unaddressed political implications of those deficits strike me as equally nightmarish. It seems a grim likelihood that, as many countries produce more and more people and less and less food, social order—and the freedom that comes with it—will not survive.

As food-growing regions of the developing world fail, people drained by the chronic hopelessness of rural poverty are laying down their farm tools and heading for the cities at astonishingly rapid rates. Failing agriculture is already swelling many of the world's great cities beyond the limits of sustainability, into megacities. In fifteen years, 27 million people will be jostling for food, living space, and water in

Bombay. At the same time, Lagos, Shanghai, Jakarta, Mexico City, Beijing, and Dhaka will swell to around 20 million each.

This is not a vague, futuristic estimate. Today a handful of megacities already contain 40 percent of all the residents of the developing world. But megacities will be turbulent, unmanageable, and violent. They are not conducive to the orderly functioning of democracy. "They are becoming monsters out of control," noted Kendall. "They threaten to outrun the control of their governments."

Uncontrolled urban growth is one reason Kendall fears so much for the future of democracy. Its consequence seems inevitable: chaos or dictatorship.

Guayaquil is the largest city in Ecuador—its principal port and one of its busiest commercial centers. This small country on the Pacific coast of South America has had a recent history of democracy and political stability. Yet the three-hour drive up the coast toward Guayaquil is heartbreaking. The highway passes mile after mile of desperate sprawl, of dilapidated houses, many of them patched with cardboard or plastic sheeting, some with only three walls, few with indoor plumbing. The nearby swampy, heavily polluted, mosquito-infested banana plantations and a cluster of pesticide factories, whose acrid odors blanket the shantytown streets on windless days, draw large numbers of Ecuadoran men down from their Andean villages in search of jobs.

(Ecuadorans had attempted to produce food through the practice of aquaculture, setting up fish farms along the coast. But pesticide-infested runoff from the banana plantations weakened the immune systems of the fish. As a result, the aquaculture effort crashed after only three years.)

At midday, the road is surrounded on both sides by clusters of people with nothing on their hands but time. As a car approaches, a boy and a girl, maybe nine, standing on either side of the highway, pull taut a piece of rope across the road. The car stops, and their father runs up to the driver's window with a shriveled banana and a small fish to sell—anything for a few pennies.

Eventually the ride leads into Guayaquil. The elegant downtown area is marked by a large night-lit Gothic church fronting on an elegant park, with wrought iron fences and lush, well-manicured greenery. The city's center is dotted with modern hotels and stores—some of whose fronts recall the elegance of old Barcelona or the glamour of Chicago's Michigan Boulevard.

The reality doesn't hit you until you go shopping. Then you see it. Just inside the entrance to every store and shop in downtown Guayaquil, a machine gun or rifle is aimed at you. Not one place of business is unguarded; the threat of poverty-driven crime is that great.

Imagine the future of democracy in Ecuador—even with its history of relatively successful civilian rule, even without the compounding pressures of global climate change. And Ecuador has a tradition of democracy.

China, by contrast, does not. No country in the developing world better embodies the potential for war, disruption, and totalitarianism than China.

Until a few years ago, Chinese agriculture was able to support most of its 1.2 billion people. Farmers maintained adequate levels of food production by drawing from aquifers, and the country imported a relatively small amount of grain. But in the last few years, many of the aquifers have collapsed or are collapsing, heavily stressed from overdrawing—and malnutrition is becoming widespread, especially in western and northern China. In 1995, for the first time in its history, China imported more grain than it grew.

As the farmlands failed, farm workers lost their jobs, and now about 120 million jobless and landless Chinese citizens—a number equal to almost half the population of the United States—are in the midst of a mass migration from the countryside to the coastal cities. There the urban population is growing by 10 percent a year, and the density of residents is three times greater than the national average.

A third of China's croplands are undergoing extreme erosion. Water shortages are becoming so severe that within the next thirty years, the entire country will reach what biologists call the "water stress" benchmark of 1,700 cubic meters of water per person. Large sections of the country will reach the "chronic water scarcity" benchmark of 1,000 cubic meters per person. Nearly half of China's five hundred major cities already lack adequate water supplies, and a quarter of them face acute shortages, according to data from the World Resources Institute and the World Bank.

For the last decade the lower reaches of the Yellow River, known as the cradle of Chinese civilization, averaged 70 dry days a year. But in 1995 it was dry for 122 days. More than a hundred large Chinese cities stagger under acute water supply problems, and only six of those meet safe drinking water standards. In China's arid northwest region, peasants must walk up to ten miles a day to secure their daily water supplies.

Given China's record on human rights, it is not hard to foresee, as the food and water crises worsen, the government detaining internal migrants on a mass basis and forcing them to labor on tiny, nutrient-depleted plots to squeeze the last possible bit of crop growth from the underwatered and overtaxed land. Another possible outcome is one for which precedents exist in the history books in an endless string of repetitions. Like its neighbor, North Korea, China may very well create a serious international disruption—if not an outright war—to divert its people into fearing some fabricated "foreign threat." I would imagine that, as China's domestic pressures escalate, national security advisers all over the world will report periodically to their presidents on their country's military readiness to engage China if the need should arise.

This scenario is one that could result from the food and water crisis—before the more severe impacts of our increasingly unstable climate make themselves felt. When those impacts are included in the scenario, an observation made by Harvard's James McCarthy takes on a new immediacy: "Had the climate change we are now beginning to see occurred 150 years ago, the world might never have been able to sustain its current population of five billion people."

In one respect the climate crisis parallels an economic crisis: it hits poor people hardest and first. "Vulnerability to climate change is systematically greater in developing countries—which in most cases are located in lower, warmer latitudes," according to Dr. Cynthia Rosenzweig, a research agronomist at NASA's Goddard Institute for Space Studies, and Dr. Daniel Hillel, professor emeritus of plant and soil sciences at the University of Massachusetts. In the already-impoverished countries of the South, they observe, "cereal grain yields are projected to decline under climate change scenarios, across the full range of projected warming." As grain supplies in the poor world dwindle, "agricultural exporters in the middle and high latitudes (such as the United States, Canada, and Australia) stand to gain" from the higher prices they can command. "Thus, countries with the lowest incomes may be the hardest hit."

Fortunately for prosperous and freedom-loving Americans and Canadians, Ecuador and China and the Middle East, with their growing food and water shortages, lie in different quadrants of the globe. North America is a blessed portion of the planet, rich with fields of wheat and plains of cattle, diverse and beautiful geographies, and an unshakable two-hundred-year tradition of democracy and personal liberty.

Can it survive the social explosion that is gathering just beyond the horizon? Today 25 million environmental refugees are roaming the world, squatting on other people's land, migrating to overcrowded cities, sneaking under cover of darkness across borders, scrabbling to survive. Their number exceeds all other types of refugees—political, economic, and religious. But North Americans don't see very many of them—as of this writing, most are located at a comfortable distance—mainly in sub-Saharan Africa, the Indian subcontinent, China, Mexico, and Central America.

In thirteen years, if not before, under current conditions, their number is expected to double. But if predictions of increasingly severe conditions—floods, droughts, storms, and temperature extremes—are realized, the tidal wave of environmental refugees will dwarf even that projection. At that point, the number could grow tenfold, to more than 200 million homeless migrants wandering the planet. This forecast is not the speculation of a chicken little. It is based on an eighteen-month research project using input from several UN agencies, the World Bank, refugee assistance groups, scientists, and field workers from all over the world.

In their extraordinarily well-ignored report titled *Environmental Exodus,* published in 1995 by the Climate Institute, Dr. Norman Myers, a visiting fellow at Oxford University, and researcher Jennifer

Kent examined areas of the world that are both the poorest and the most vulnerable to the ravages of climate change.

Assuming that crop yields continue at 1985 levels:

- In the next thirteen years sub-Saharan Africa's already stressed food production will decline by 20 percent—leaving more than 300 million people in a state of permanent malnutrition. (Even as I write, in June 1996, a story on the news wire confirms the trend. The UN Food and Agriculture Organization (FAO) is reporting that 22 million people in sub-Saharan Africa are facing food emergencies. Tight global cereal supplies, high prices, the serious balance-of-payments difficulties faced by many African countries, and low availability of food aid "threaten to undermine sub-Saharan Africa's food security" the FAO report adds.)

- Changes in the monsoon patterns that provide India with 70 percent of its rainfall will inflict severe shortages and disruptions on the country's projected population of 1,190,000. Even a half-degree Celsius temperature rise will reduce the wheat crop at least 10 percent.

  To put this in perspective, recall that a half-degree Celsius increase is far below the lower limit projected by the IPCC. Skeptic scientists have called such a rise in temperature "negligible," but given its potential to cut into India's food supplies, I would be surprised if many Indians agreed with that disembodied characterization.

- In thirteen years 3 billion people—more than half the population of the developing world—could be cutting down trees for fuel and firewood. As displaced peasants slash and burn their already-depleted tropical forests, they will further reduce the capacity of the planet to absorb carbon dioxide and will release it instead into the atmosphere, where it will accelerate the pace of warming.

- Disease outbreaks, driven by changing climate patterns,

> will parallel the spread of poverty and displacement. As refugees cut trees and burn grasslands to make new settlements, they will unleash rare or remote microorganisms, infecting themselves and contaminating others. Even without Ebolas, climatic instability is fueling the recurrence of age-old diseases in the poor areas of the world. A 1994 outbreak of plague in India was directly caused by climate extremes. That summer blistering temperatures of over 120 degrees Fahrenheit, combined with an unusually long season of monsoon rains, produced exceptionally high levels of humidity. That humidity created the breeding conditions for fleas in grain-storage centers. Later that summer, flooding from the unseasonal rains spread garbage through the cities, which triggered an explosion of rats. The fleas attached themselves to the rats, then spread the plague to humans, hundreds of whom died of its effects. Subsequently, flood-driven outbreaks of malaria and dengue fever took place in parts of the country where they had never before occurred.

Long before sea levels rise by two to three feet over the next century—and they are projected to continue to rise after that—the flood of environmental refugees will likely have overwhelmed both our compassion and our capacity to help. Bangladesh has lost 600,000 people to cyclones and storm surges over the past thirty years. In the next fifty years, given its vulnerability to coastal destruction and inland flooding, Bangladesh could, according to conservative estimates, contribute 26 million more refugees to an increasingly homeless world, according to Myers and Kent. Another 12 million refugees would be driven from the flooded Egyptian delta. Anywhere from 70 to 100 million Chinese citizens would be forced to migrate to new habitats, and 20 million residents of India would be displaced by the compounding effects of climate instability.

Where does democracy fit in this picture? Can there be any doubt that within the next century the migration of some 200 million environmental refugees will force many governments into martial rule?

Even in the United States, with its geographical variety and natural abundance, climate change could have explosive political repercussions. Climate change, for one thing, alters the distribution of rainfall. As a result, the continental interiors that produce the grain that feeds much of the world may well experience recurrent and increasingly severe droughts.

In 1988, the hottest year on record to that date, a midwestern drought depressed grain yields by 30 percent, dropping U.S. crop production below consumption requirements for the first time in three hundred years. The same year, Canadian grain production dropped by 37 percent.

The interrelated web of the global environment includes even the fertile plains of North America. "Unexpected 'surprises' may well accompany the buildup of greenhouse gases," according to Rosenzweig and Hillel. . . . "Under changing climate conditions, farmers' past experience will be a less reliable predictor of what is to come. . . . A seemingly small change in one variable—for example, rainfall—may trigger a major unsuspected change in another; for example, droughts or floods might possibly disrupt the transport of grain on rivers. One 'surprise' may then lead to another in a cascade, since biophysical and social systems are interconnected."

In 1995 scientists studying the natural history of our own great plains arrived at an unnerving finding. They discovered that the wheatfields could turn—in a decade—into a vast desert. As recently as the turn of the century, a prolonged drought launched a migration of sand dunes across the face of Kansas and eastern Colorado. The discovery brought home the fragility of the ecosystems that provide our food. The climate period we live in—scientists call it the Holocene—is apparently not as stable as we once thought. Small changes in the climate can lead to large changes in ecosystems.

This is what is so frustrating about the arguments of the oil and coal lobbies, and especially their skeptic scientists. Their minds must work overtime to contrive ways to deny so fundamental a reality: There are far more interconnections between our physical environment and our social existence than we will ever be able to identify, let alone control.

In a perverse way, many of them are contradicting their own political positions. They advocate expanded freedoms and a drastic reduction in the power of government to control its citizens. Yet in their blind presumptuousness, they believe they will be able to minimize and control people's reactions to the disturbed forces of nature.

They tell us we can solve our problems through more economic growth, but they ignore the fact that such growth has impacts on the physical world. They tell us, more specifically, that any agricultural problems that are likely to arise will be taken care of by the self-correcting forces of the marketplace, but they refuse to acknowledge that the very farms on which the agricultural marketplace depends are vulnerable to the ravages of climatic change.

The Greening of Planet Earth, the well-circulated 1991 coal-industry propaganda video, touts the abundant benefits of enhanced carbon dioxide on agriculture. But Rosenzweig and Hillel consider the idea that the agricultural market can remedy the consequences of climate change to be particularly destructive and simplistic. One "notion which can be . . . misleading," they write, "is a blind faith in agriculture as a self-correcting process: that through forces of the market and self-preservation farmers can and will readily and fully adapt to climate change as it occurs. . . . They will certainly make every effort to do so, but the efforts of farmers may be constrained or even thwarted by factors beyond their control."

That misleading notion—"along with the convenient expectation by some plant scientists that the physiological effects of enhanced CO_2 will be overwhelmingly positive—may lull decision makers and the public at large into complacency regarding global warming. . . . Global warming is, in our opinion, a real phenomenon that is likely to engender serious consequences."

For the sake of argument, let's retreat into a more comforting scenario. Let's assume that the droughts in the wheat-growing areas of North America are not so severe that they starve our own population. They only starve those in the poor countries who depend on us for their basic nutrition. It is a scenario Argentine diplomat Raul Estrada Ayala calls "a green North and a brown South."

The greenery of such a North would be deceptive. It would

conceal a political and moral time bomb. It is hard to imagine that a society that fortresses itself against the rest of the world could continue to be an open society, vibrant with freedom, productively democratic, peaceful, and secure. It is hard to imagine that it would be able to nurture those human qualities we prize—hard work, ingenuity, compassion, and intelligence.

Environmental disruptions in the poor areas of the globe will not remain conveniently compartmentalized within their borders. If displaced refugees in South America, Asia, and Africa continue to burn trees and grasslands for fuel and settlements, that removal of vegetation will accelerate global warming. The plants and trees of the terrestrial ecosystem are the largest absorbers of carbon dioxide, which otherwise rises into the atmosphere. Nor is it the environment alone that overflows national borders. The economy is also global. As more and more inhabitants of the poor countries are displaced, the emerging markets of the developing world will begin to collapse—exerting a tremendous downward pressure on centers of trade, finance, and manufacturing in the North. Without the continued development of emerging markets, the international economy will begin to contract, severely eroding the basis of its survival.

The United States, like any open society, is vulnerable to terrorism. A significant surge in terrorism is the likeliest result of the desperation that is overtaking many people in environmentally disrupted countries. "The World Trade Center was easy," Norman Myers says. "The next time a nuclear device is set off, it most likely will not be by a government. It will probably be set off by some group of people who are so frustrated at being consigned to desperation that they will be driven to potentially outrageous acts of terrorism."

Both the Congress and President Clinton share concerns about terrorism. In 1996 they collaborated to pass an antiterrorism bill that extends the surveillance capabilities of the FBI into a new—and arguably unconstitutional—realm. In the name of preventing terrorism, the legislation extends the Bureau's intrusive powers. It seems only a matter of time before similar powers are extended to the Immigration and Naturalization Service.

The 1996 antiterrorism bill was deeply offensive to Americans on both the left and the right. But in the wake of the Oklahoma City bombing, politicians played on the insecurities of a stunned and deeply saddened country to move the country a step closer to a police state. Still, the bombings in Oklahoma City and Centennial Park during the Olympics in Atlanta may pale before the violent social upheavals that will be triggered by climate change.

How democratic can we expect our government to be in the face of the weather extremes that lie just ahead and that will overtax its ability to provide relief: a series of record-breaking midwestern floods, a succession of Hurricane Andrews that tear hunks out of south Florida, too many too-dry crop-wilting Julys and Augusts in Nebraska and Kansas, a revolt of jobless fisherman that spreads up the West Coast from San Diego to Puget Sound?

We have already entered a period of unusual political volatility, but that volatility can only increase in the face of weather-related catastrophes in the coming years. The economic impacts alone will be considerable. Increasingly stressful disasters will exhaust relief budgets. Disease-producing temperature extremes will lead to a profusion of sick days, which will in turn detract from our economic productivity and add unanticipated costs to our already-stressed health insurance system. The foreign markets we are cultivating to expand our economy will shrink. Some of our manufacturing industries could be crippled by the disappearance, due to localized climate disruptions, of foreign commodities critical to their production processes.

When you think it through, institutional repression is the most probable political outcome of climate change—in the United States almost as much as in the flood-prone coastal countries in the developing world.

One response might be the militarization of the Federal Emergency Management Agency (FEMA), which will be called on to respond to these kinds of crises. The agency has already pursued a strong set of flood control and response programs. But for now, FEMA has still not been authorized by Congress to begin to plan for disastrous climate change impacts. FEMA officials say they cannot consider the broader impacts of climate change unless Congress includes it in its legislative mandate.

Another governmental response might be food rationing—which would inevitably be accompanied by black market crime. One can also imagine, if major roadways, railbeds, and other parts of the infrastructure are disabled by weather extremes, vastly expanded state and regional police forces. Given the likely social consequences of climate disasters, one can most easily imagine a quantum leap in the size and intrusiveness of federal internal security forces. No dissenting group of people—political, environmental, or religious—will avoid their surveillance and, likely, their infiltration.

It is easy to comprehend the threat of official repression by government agencies. History is full of examples of repressive governments. The lessons of the past tell us what to expect.

Listen to William Ruckelshaus, today a prominent captain of industry who knows the inner workings of big business. But as the first EPA administrator and as a delegate to the 1987 World Commission on Environment and Development, he is also keenly aware of the threats to the global environment. (As one of the Watergate-related casualties of President Nixon's "Saturday Night massacre," he was willing to sacrifice his professional career for the sake of democratic principles.)

"This is democracy's time on the world stage," he said in an interview. "With the collapse of authoritarian systems in the Communist world—which proved even less capable than democracies in dealing with global warming and other chronic environmental problems—this is our chance to show we're capable of coping with our own complexities.

"If we do not bring ourselves to solve these chronic problems, all of our free institutions will be challenged. The institutions themselves will be the first things to go. I don't honestly believe that global warming can't be solved. Not that we can turn on a dime, but we can and should set in motion new policies that will come to fruition over the next several decades.

"The real question is whether we can deal with them in the context of freedom. It is a very open question. We could well be forced to revert to some sort of dictatorship to act in the face of the kind of problems posed by the global climate."

If governmental repression is at least comprehensible, however, it is harder to visualize the more subtle threat to our freedoms that might be imposed by the largest and most powerful industry in history—the one most threatened by the gathering climate crisis.

To date, the fossil fuel industry has kept the issue of climate change out of public consciousness—with expenditures that amount to mere corporate pocket money. What would it do in the face of a more imminent threat to its survival?

I surely don't mean to suggest that corporations are eager to abuse citizens' civil liberties. They are not in the business of government. No long-range corporate strategic plans include repression. But when their basic survival becomes threatened, that situation could change in ways we are not prepared for. Certainly they will lean with all their corporate weight on the government to protect their interests. But whatever the government can't do, they will try to do themselves. Survival is the motive force of any enterprise—especially those that have been tempered in the crucible of the ruthlessly competitive late-twentieth-century global economy.

Private and corporate security forces have already grown significantly over the last twenty years. In pursuit or protection of profits, companies have intercepted mail, tapped telephones, and conducted extensive surveillance of adversaries. Corporate whistle-blowers and grassroots community groups have endured harassment. An international information war is being waged over trade secrets and has become so pervasive, it is now a major focus of the CIA.

The early round of attacks by the fossil fuel industry on established climate science provides one clue to how the oil and coal lobbies might behave in the future—and to the fragility of the public's right to know. Personally, I find the industry's assaults on the integrity of the scientists—in collaboration with its ideological allies in Congress—even more intimidating.

Like most power grabs, that of the oil and coal industries begins at the relatively invisible level of determining public perception—waging war through the media to suppress areas of reality that are likely to become the most active arenas of conflict. Their campaign is a prelude to a kind of totalitarianism that is based on the control of information and the subversion of truth. The core technique is to turn truth on its head—and then relentlessly repeat the untruth. It recalls the society of George Orwell's *1984,* in which citizens are constantly bombarded with the message: "War is Peace. Slavery is Freedom. Ignorance is Strength." In this case, however, it is not the dictatorship of a government that looms. It is a dictatorship of corporate interests, whose well-rehearsed attack squads assault scientific truth and personal reputation with equal ruthlessness.

For people in the developing world, the onset of martial law may lie just beyond the next wave of climate-triggered disasters. For those of us in the wealthier countries, relatively unaffected as yet by climate change, it may lie just beyond the next pulse of propaganda. Overlay the politically volatile times we live in with a succession of climate-induced body blows to the nation's economy, and the highest level of personal freedom in all of human history may well become a fond memory.

This much is true of totalitarianism: it usually begins with the lies.

In the late spring of 1996, while much of the Midwest was reeling from unseasonal snowstorms and floods, parts of Kansas and Oklahoma were suffering from the worst drought in a century. The drought destroyed millions of acres of wheat, triggering a 12 percent wheat shortage—the smallest harvest in 16 years. The two successive poor growing seasons, combined with increasing international demand, have dropped U.S. wheat reserves to their lowest level in 50 years. Recalling the debilitating Dust Bowl of the 1930s, 71-year-old Lewis Mayer told a New York Times *reporter that he had decided to plow under his winter wheat crop, which he had declared a total loss. "There's just no rain at all. It's hotter than blazes," Mayer said of his Oklahoma farm. "I was a ten-year-old boy in 1935, and as near as I can remember, it just looks similar. Very similar." The impact of the drought of 1996 on farmers in the Southwest, said Texas agricultural commissioner Rick Perry, "has the potential to have the magnitude of nothing we have seen in history. It is a devastating thing, mentally and spiritually."*

On November 8, 1996, the worst cyclone of the century killed nearly 2,000 residents of the Hyderabad section of India.

The storm, which was fueled by warm surface waters in the Bay of Bengal, destroyed thousands of homes, ravaged crops and livestock, and triggered an outbreak of cholera. "The most beautiful and fertile rice-growing district of [the region] has turned into a burial ground," Chandrababu Naidu, an official of Andhra Pradesh, told reporters. Some 500,000 people were stranded without food by the storm, according to officials. The storm swept away tons of rice and turned banana and sugar cane plantations into swamps. Inundated fields were littered with the decaying corpses of livestock. When the cyclone struck, the region was still recovering from another major storm, three weeks earlier, that had killed 350 people.

EIGHT

One Pathway to a Future

*THROUGHOUT HISTORY, IT HAS BEEN PHILOSOPHERS, RELI-*gious leaders, and revolutionaries who have asked us to reexamine our values, our relationships, our purposes, and the way we live. Now we are being asked by the oceans.

At first consideration, it is a challenge of paralyzing proportions. The complexities of the global economy are almost as intricate as the complexities of the earth's ecosystems. The possibilities for disruption and radical change evoke visions of mass poverty and social chaos. I don't think these visions need be the inevitable consequences of economic change, but I am not certain of this. It doesn't matter. I don't believe we have a choice. What is far more certain are the consequences of inaction.

The pressures militating against change are pervasive—and denial is a formidable obstacle. We are afraid, on one hand, of the impacts of global climate change. The climate snaps that the ice cores tell about are truly horrendous to contemplate. The prospect of the world becoming a storm-battered, insect-infested breeding ground of infectious disease, of temperature extremes, of extensive drought and desperate heat, is truly terrifying.

On the other hand, our existing energy system connects to every corner of the economy. It underlies and supports our twentieth-century lifestyles. After food and shelter, energy is our most basic need. The prospect of tampering with the elaborate structure of resources and institutions that provide us with energy is in itself overwhelming. If the scale of the problem is paralyzing, so is the magnitude of change required to solve it.

The issue of global warming first attracted national attention in the middle of the 1980s. Since then it has followed a strange dolphin-leaping cycle, surfacing into public consciousness, only to recede once again. The earlier scientific uncertainties explain this pattern only partially. Our own reflexes of denial explain it more completely. We let it in a little bit, then close our eyes. When it forces itself back into our consciousness, we shut it out again.

It is traumatic, this understanding, if we let it be. It is desolation.

The experience of Dr. Daniel Goodenough, of the Harvard Medical School, provides an illustration. Recently, the medical school mounted a series of three seminars on human health and the global environment. The first seminar was overflowing with energetic, aware, and concerned young medical students. Inexplicably, however, the second session was only half full. At the final seminar only half a dozen students showed up.

When the perplexed Goodenough asked the students why attendance had fallen off so sharply, their responses were identical. The material was compelling, they said, but it engendered overwhelming personal reactions. The problems were so great—and the ability of the students to affect them so remote—that they could deal with their feelings of frustration and helplessness and depression only by staying away.

News stories about the warming of the planet generally evoke an eerie silence. Several front-page stories have appeared in *The New York Times,* as well as recent cover stories in *Time, Newsweek,* and *Harper's,* and a flurry of other highly visible media sources. But to my frustration, it is clear that this is a story that people do not want to hear.

That could change very quickly. American politics are marked by extraordinary volatility. Seismic shifts like the fall of Communism and the end of apartheid happened with stunning suddenness. Over the past four years the preferences of the voting public have swung wildly from Bill Clinton to Newt Gingrich to Colin Powell. The pervasive denial of global warming that so frustrates the reporter in me could perhaps change with equal suddenness. We all see the signs. The extraordinary weather extremes of the past few years evoke intuitions that are increasingly difficult to ignore.

We owe today's young people more than overwhelming information. We owe them a solution. Against the backdrop of escalating climate change, emotional denial is an indulgence we cannot afford. We need to act quickly and decisively, and we need to act on a large scale, one that far transcends the partial and inadequate measures now being considered at the UN climate negotiations. If we do not, I am afraid that our societies will deteriorate into either chaos or totalitarianism.

The global climate seems to have issued that ultimatum. Humanity would not be made extinct by a surge in climatic instability, but the intricate fabric of interrelationships that constitute society would be ravaged in proportion to the magnitude of the disruptions. Philosophically, such a blow to our highly complex institutions—and a blow that is self-induced, no less—would mean that everything our civilization has accomplished to this point would become basically meaningless.

It would be a judgment on our adaptive capabilities as a species—that we have exhausted our intelligence and creativity and have instead become a collective infection on the planet.

It would mean that all of human love and creativity and accomplishment—from the building of cities to the creation of music to the thrill of discovery—will have amounted, in the final balance, to nothing more than a hollow ego-inflated illusion of achievement. And it would lead to our descent into an earlier, more brutish state. Losing the capacity to appreciate the world around us, unable to discern and understand its meanings, we would become more defensive, more predatory, more ignorant than what we have worked so hard all these centuries to become.

That awareness should move us to begin acting now. That, and the knowledge that we have an extraordinary opportunity to make right the conditions of the planet and to heal some of the deepest wounds that divide humanity.

The fossil fuel lobbyists aside, virtually all experts agree on the need to stabilize the increasingly turbulent global climate. At the simplest level, that will require a change in our energy technologies. It will require us to shift from coal and oil and, eventually, from the lighter-carboned natural gas, to renewable energy sources that do not generate heat-trapping greenhouse gases. Those technologies exist today.

At another level, we must address the massive differential in living standards between rich and poor—what writer Tom Athanasiou calls the divided planet—if only to protect ourselves against the coming pulse of carbon from the developing world. Extending clean energy technologies throughout the world would provide the starting point for addressing this problem as well.

At a level only slightly more abstract, we also need to change the way we calculate environmental costs. Today, when we use up a resource—for instance, when we overfish the coastal waters of the North Atlantic—we count the product only as profit. The corresponding extinction of the fish is nowhere reflected on our balance sheets. Similarly, when we generate pollution—which shortens lives, fouls water, and reduces the growing capacities of the land—we must reflect those costs on our economic ledgers. Without a more environmentally complete account of our economic behavior, as economist Herman Daly has pointed out, our systems of economic measurement will bear no relationship to the capacities of the biosphere to support us—and we will continue to destroy our foundations, even as we deceive ourselves that we are reaping ever-greater financial dividends.

The final change we need to make lies at the level of morality, both personal and collective. This is where the purposes of our existence are defined—the level of our most basic self-identity.

For most of the thousands of years of human existence, people have thought of themselves as dependent children of nature—subject to her whims and tantrums and blessings. Human communities were so small and the world was so big that they could be destroyed by her floods and storms, supported by her seasons of abundance, and brought down by her times of scarcity. As a society, we continue to think of ourselves that way. But in so doing, we ignore the looming but largely invisible truth that, somewhere in the recent past, our collective power to alter the environment has become as great as that of the basic systems of the earth.

With the growth of the human population to five plus billion and with the development of extraordinarily powerful technologies, we ourselves have become a force of geological magnitude. Our technologies of transportation have opened every corner of the world to human exploration and settlement. Our technologies of communication have shrunk the globe, joining people from every land in a global network of instant communication and ongoing dialogue. And the massive emissions from our factories, cars, and homes are altering the very seasons of the earth and the balance of its most fundamental cycles of life.

It is not only the carbon cycle that we are altering. We may also be altering the central biological process of evolution. In his Pulitzer prize–winning book *The Beak of the Finch,* Jonathan Weiner reflects on the El Niño event of 1982–83, one of the strongest of the century, and its effects on the evolution of finches on the Galápagos island of Daphne Major. "It is no exaggeration," he remarks, "to say that a lasting change in the ocean currents—especially a change in the intensity or frequency of El Niño—would change the course of evolution in Darwin's islands. . . . Before [1982], Darwinian selection was keeping Darwin's finches apart and distinct. After [that year] selection pressures on Daphne began forcing the birds together. If global warming does bring more extreme Niños, then the selection pressures on the islands may take a very long time to return to what they were before. . . . In other words, in the present climate of the Galapagos, it would take a thousand years of not unlikely weather to create a new species of Darwin's finches on the islands. [But] if the climate were to change and inflict a series of grim droughts or floods at just the right intervals, without missing a beat, it could create a new species in a single century. . . . The case suggests the kinds . . . [of] unpredictable local events that global warming may inflict in our lifetimes, even on some of the most isolated islands in the world."

If our industrial activities are reversing the planet's carbon cycle and altering the rate of evolution, I think it is time to rethink who we are. It is time for us to look into the mirror and acknowledge that we are no longer helpless, dependent children of nature. It is time to finally acknowledge the power and the consequences of our activities.

It would nice if we had time to grow into the realization of who we have become and bring ourselves gradually to address our adult responsibilities for the creative and destructive forces we exert. But what is so unfair to all of us is that we do not have that time. We have been blindsided by the stunning speed of our recent growth.

A hundred years ago a Swedish chemist named Svante Arrhenius discovered that a buildup of atmospheric carbon dioxide could lead to planetary warming. But global warming has been recognized as an imminent threat only within the last ten years. We have only an incredibly short time in which to change our self-image as a species—yet it is a self-image that has surely been a part of our thinking since the dawn of human consciousness.

Regardless of how you feel about contraception, abortion, and laboratory fertilization, about assisted suicide and artificial life support, the very fact that they are issues of the day reflects our newly acquired power over the beginnings and endings of life. Our job now is to learn to use that power in ways that make us feel right about ourselves as human beings. Our earth demands the same ethical maturity from us. The signals of its climate tell us it is time we stopped ducking our responsibilities.

Today diplomats are haggling over conflicting proposals, none of which, if adopted, would do much to head off the increasing turbulence of the global climate. The best proposals call for reducing our emissions of greenhouse gases by 20 percent below 1990 levels. But scientists tell us that if we want to stabilize atmospheric conditions at relatively comfortable climate levels, we must cut those emissions by more than 60 percent.

The danger is that policymakers will decide on an arbitrary, politically attainable target for emissions reduction, after which they will assume they have addressed the problem of climate change. The truth is that regardless of what target is negotiated by politicians, any significant human alteration of massive planetary systems will have potentially serious effects on the global climate.

The industry-sponsored skeptics have made a living off of scientific uncertainty. They are right about one thing. Scientists today do not know whether there are thresholds of warming that will propel us into greater magnitudes of climate change—or even how the climate instabilities we are already seeing are likely to play out. Nor do they know whether it is already too late—regardless of what we do—to prevent the disruption of civilization by climate change.

All they can tell us is that climate change will not be a linear, gradual process. Judging by the record of the ice cores and by the dynamics of other natural phenomena, surprises are indeed likely to occur. What that tells us, I think, is that we need to move quickly—even as we keep our fingers crossed and hope the escalating, record-breaking weather extremes don't become too disruptive.

But if the science still contains uncertainties, our behavior does not.

Here are some of the highlights of 1995, as recounted by the Worldwatch Institute:

- Despite global agreements to limit carbon emissions and a voluntary U.S. plan instituted by President Clinton, a record 6.1 billion tons of fossil fuels were burned that year.
- The world's grain harvest was the smallest since 1988, and the grain reserves—the grain available to the world for consumption if all production stopped—reached an all-time low of just forty-eight days.
- Insurance industry payouts for weather-related damages reached $57 billion for the 1990s—compared with only $17 billion for all the 1980s.

The climate crisis is no longer a question of science. The fact that the future rate of warming has not yet been established—or that all the various geographical impacts have not yet been identified—is, from the point of view of social policy, irrelevant. The science has already told us all we need to know to begin to act.

We already have available the climate-friendly energy technologies we need to stop inflaming the atmosphere. What most non-polluting energy technologies need is not more research and development, but an infusion of capital to make them as inexpensive as coal and oil. Economies of scale would result from their world-wide mass production, and through the natural processes of industrial evolution, the next—and more productive—generation of those technologies would soon emerge.

As energy analysts Joseph Romm and Charles Curtis of the U.S. Department of Energy have pointed out:

- Today, with new technologies of enhanced fermentation, the cost of ethanol has dropped from $3.60 a gallon to $1. In a few years fuel could be produced from the sludge of our municipal sewage systems, from crops and crop waste, and from wastepaper for 60 cents a gallon.

- A technology that uses rooftop panels to convert sunlight directly into electricity now costs one-tenth of what it cost twenty years ago. It is so promising that Royal Dutch/Shell recently bought two photovoltaics (PV) companies. Photovoltaic technology could provide electricity to some 2 billion rural residents of the developing world. In addition, thermal solar energy—which uses the heat of the sun directly to warm water for home heating and hot water use—has virtually untapped energy potential.

- The technology that many specialists think is the most promising replacement for coal and oil involves the conversion of hydrogen into electricity. The devices, called fuel cells, were developed by NASA for use on space missions; they could power homes, factories, and cars with no greenhouse gas emissions and virtually no pollution. Unlike the combustion involved in burning coal, oil, or natural gas, fuel cells create electricity essentially by reversing the pro-

> cess of electrolysis that splits water into oxygen and hydrogen. Small, silent, and efficient, they can be installed in the basement of a building. Their waste heat can be used for space or water heating, and their only nonpolluting by-product is water.

Christopher Flavin of the Worldwatch Institute recently reported that:

- The fastest-growing energy market is for wind power, which in 1995 produced 1,500 megawatts of electricity.
- The cheapest, most plentiful, and cleanest fuel available today—hydrogen gas—is already being used commercially. Mazda and Mercedes have already developed hydrogen-fueled cars. All the world's current energy needs could be met by the hydrogen extracted from less than one percent of the global supply of fresh water. It could be distributed by the same pipelines we currently use to carry natural gas.

During a recent presentation at Tufts University, longtime alternative energy specialist Amory Lovins explained the design for a new nonpolluting "hypercar" that has already attracted more than $100 million in investments from private companies. The car, which is built of lightweight carbon fiber, can be as big and fast and safe as any conventional car on the road today. It conceptually resembles a computer on wheels, Lovins said, more that a traditional internal combustion auto. It is powered by a hybrid engine fueled either by hydrogen or by electricity from a fuel cell—along with a flywheel-type mechanism that stores energy during acceleration and releases it for fuelless propulsion.

In addition, proven energy-efficient techniques could be put to greater use—superinsulated buildings, heat pumps, gas-fired cogeneration plants that reuse their waste heat, computer-controlled lighting, and highly efficient heating systems. With their dramatic fuel reductions, these techniques are already saving huge quantities of money and energy without impairing living standards.

As a bridge to a new energy era, the economics panel of the IPCC has identified a number of steps, called "no regrets" policies, that at virtually no cost at all could reduce greenhouse emissions by around 20 percent. They include such simple steps as implementing known efficiency and conservation techniques, planting more trees (to absorb carbon dioxide), and instituting international standards for energy-efficient appliances.

These steps are no-brainers. Even if there were no climate threat, their implementation—which could be done at little or no cost to society—would improve air quality, protect drinking water, enhance agricultural productivity, and lower energy costs. Put to use by themselves, they are not enough to head off atmospheric warming. But during, say, a ten-year energy transition period, they could be a crucial part of a much larger transformation.

A far more drastic element of any energy transition must be the transfer of all the millions of dollars in government subsidies and tax incentives away from fossil fuels and into renewable energy. This step would elicit a storm of counterattacks from big oil and big coal, yet it would be critical.

The federal government spends more than $20 billion a year to subsidize the development of oil, coal, and natural gas. It spends more than $10 billion to subsidize nuclear energy. According to energy economist Doug Koplow, the current "pattern of subsidies does little to promote cleaner energy technologies or to protect the environment. Yet the patterns have persisted . . . due to the substantial political power of the recipient sectors and to a lack of timely, standardized information on the scope and magnitude of the subsidies."

It is time for the government to stop subsidizing the destruction of the global environment. Diverting these tax credits and subsidies to the commercialization of windmill farms and home-based fuel cells and photovoltaic panels and hydrogen fuel plants would provide the liftoff boost to propel renewable energy into the big league of global industry.

What needs to happen in the long term is not hard to understand. But it is an undertaking without any precedent. We have no blueprint. Still, it helps to remember what Alden Meyer and other energy specialists point out: In every historical energy transition that humanity has undergone—from wood to coal, from coal to oil, from oil to gas—each new energy source has been cleaner and far more efficient than the one before.

The great news is that each historical energy transition has resulted in an explosion of economic progress. Although the fossil fuel industry would like us to believe otherwise—that shifting away from coal and oil would mean jettisoning our current standard of living—a new, deliberately managed energy transition could actually benefit us all in a very big way.

The oil lobby argues, for example, that you can't deprive Americans of the cars they love. And they're absolutely right. But even the most car-loving American doesn't care much whether his auto runs on gasoline or on hydrogen. Existing pipelines could carry hydrogen gas just as well as natural gas. The argument that an energy transition would constitute a threat to our freedom and standard of living deliberately confuses the fossil fuel economy with the larger economy of which it is only a part.

I can understand why the oil and coal lobbies paint such dire, doom-and-gloom scenarios of the consequences of eliminating our use of fossil fuels. They want us to be frightened about our economic security. Our fear is their best weapon.

The Business Council for a Sustainable Energy Future, which represents alternative energy producers, has its own economic agenda. Still, I find its director, Michael Marvin, persuasive when he says we should not fear leaving a carbon-based economy. "It is not as hard as it might seem," he says, "nor will it be nearly as disruptive to the economy as they would have you believe."

That assessment depends, in part, on measuring the economy in such a way as to incorporate environmental considerations. As the economist Herman Daly has said, "There is no point of contact between our current measurement of economics and the environment." By ignoring the costs of pollution and failing to count the loss of dwindling resources, traditional accounting treats the earth as "a business in liquidation."

In 1995 three economists asked the question: "If the economy is up, why is America down?" The reason they gave is that the picture of our economic health provided by our traditional economic measuring stick, the Gross Domestic Product (GDP), is highly distorted. The current system measures the quantities of economic transactions regardless of their social and environmental impacts or anything else, and it fails to register whether those impacts are negative or positive. Hence "the Gross Domestic Product portrays disaster as gain," the authors argue. "The nation's economic hero is a terminal cancer patient who is going through a costly divorce. The happiest event is an earthquake or a hurricane. The most desirable habitat is a multibillion-dollar Superfund site. All these add to the GDP because they cause money to change hands."

The three economists, members of a nonprofit institute called Redefining Progress, as well as Daly, would not only like to implement environmental accounting, but would also like to extend the idea into the realm of social benefits and deficits. But I think that carries economic measurements into areas that are likely to be subjective and difficult to quantify, and would generate endless arguments for which we do not have time.

The costs of resource depletion and environmental damage, however, are more easily defined. If the yardstick of our economic health were altered to include them, the extraordinary costs of coal and oil burning to both the planet and to the global economy would become graphically apparent.

But ultimately the global climate crisis demands far more than "no regrets" policies and accounting changes. It demands an even bigger leap. It demands a crash program like the Manhattan Project.

At the start of the second World War, the atomic bomb existed only in theory, in scientific equations. The Manhattan Project brought together the best nuclear scientists and engineers in the United States to create a real atomic bomb. They accomplished this feat in the astoundingly short time of two and a half years. It is testimony to the extraordinary things people can accomplish under pressure when they believe their security is threatened and when they have the unified support of their fellow citizens.

What the climate signals are telling us now is that we need a similar effort to replace all our coal- and oil-fired engines of production, transportation, electricity-generation, and heating with renewable energy systems that do not release destabilizing gases into the atmosphere.

Contrary to the coal and oil industries, the issue is not whether we can afford such an effort. Nor is the issue whether we can actually do it. The only issue, I believe, is how.

Unlike many other environmental problems, the climate crisis is not one that is affected by the actions of individuals—by giving up extravagant personal habits and shifting to recycling or car pooling. The solutions, I think, need to happen at the level of institutions, and then they will reverberate outward through society.

In this postindustrial age, national and transnational corporations have grown to mammoth size because of the human demand for mass-produced goods and the historically sudden interconnectedness of the world. These corporations are, in most cases, extraordinarily productive engines of creativity—not only in the goods they produce but in the ideas and innovations they generate. Chemical manufacturers, in league with governments, have already succeeded in halting production of ozone-destroying chemicals. It is not surprising that they would now want to oversee the transition to zero-emissions energy sources. Unlike the ferociously resistant oil and coal industry, with all its weapons of denial, all the other corporate players in the climate negotiations basically accept the existence of the gathering climate threat. Most of them agree on the need for an energy transition with enforceable timetables.

The International Climate Change Partnership of large nonoil corporations wants the process to be designed and controlled by corporations—since it is corporations, they argue, that know the processes of manufacturing and distribution and that can accomplish the transition with the least possible disruption to the global economy.

While I applaud their concern, I think their prescription is wrong. I severely doubt the corporations, on their own, would ever get it done—and surely not in time to avert major disruptions to our collective existence.

For one thing, the ICCP proposal assumes good-faith compliance by the coal and oil industries.

Even in the unlikely event that that compliance is forthcoming, however, I do not believe that the captains of industry are capable of discerning the proper balance between the needs of the planet and those of the economic order. All their training, all their orientation, all their professional motivation has been directed toward improving corporate profitability. They know how to engineer efficiencies into their production processes, how to expand their marketing to increasing numbers of global consumers, and how most profitably—if not most humanely—to manage their workers.

But they are not trained in altruism. Without a system of enforceable regulations, it would be extraordinarily difficult for them to sacrifice their competitive position—even for the earth's well-being. It would be impossible, I believe, for them to keep their eye on the bottom line of profitability and at the same time on the upper reaches of the carbon-burdened sky.

Unfortunately, the vision of too many executives is limited by the conclusion of the economists who trained them: Adapting to climate change is too expensive. They don't understand what economics journalist Robert Kuttner calls the limits to markets—the failure of our economic system to account for, among other things, damage to the global environmental commons.

Executives' fears about cost reveal a short-circuit in their thinking. They don't understand that the global environment circumscribes the global economy and that they cannot negotiate rates of economic growth and carbon dioxide emissions with the biosphere. They fail to appreciate that if the costs of changing our energy diet are too high, the costs of *not* changing it will be incalculably higher.

Here is what is so simple—and yet apparently beyond their grasp:

The laws of supply and demand do not supersede the laws of nature—and when those two sets collide, the physical planet is the court of highest appeal.

Then there is the fatal lure of the newly globalized economy. Any solution to the climate crisis depends on restoring some measure of equity to the vast majority of poor nations. Without at least a partial closure of the economic gap—specifically, through the transfer of climate-friendly energy resources—the heating of the planet will one way or the other undo us all.

Most of the business community, however, sees the climate issue as yet another opportunity to sell yet another category of goods—in this case, renewable energy technologies—to developing countries. That corporate view of India, China, Mexico, Brazil, and Eastern Europe as markets for exports is, I am afraid, lethal to our hopes for staving off a climate crisis. Those countries can barely afford to feed their people, let alone finance energy transitions.

Remember that this assessment assumes the good faith of the fossil fuel industry. Unfortunately, there is not much basis for making that assumption. With the exception of a few companies like Shell, or at least their public utterances, the fossil fuel industry has worked as hard as it can to obscure the scientific confirmations of global warming, to obstruct attempts at international agreement, and to deny a problem of cosmic proportions.

For the millions of dollars they have spent on public relations, all they are getting is a strategy so stupid, it will end up undoing their industry. They are villainizing themselves unnecessarily. The leaders of the coal and oil industries did not plan the climate crisis. They did not deliberately reverse the carbon cycle and force the oceans to rise. The scope of the problem dawned on all of us only a handful of years ago.

But like other business leaders, fossil fuel-industry executives have become largely economic beings. Even their most personal thinking is informed by their economic environment. And that makes me worry that if we let them, they will take us all down before they let go.

In the course of writing this book, I interviewed many people. At the end of most of my interviews, I asked all of them—scientists, politicians, environmentalists, diplomats, business leaders—the same question: If you were king of the world, what would you do? Their answers ranged from using their power of government to set an example by installing energy-efficient light bulbs and powering buildings with renewable energies, to eradicating greed from human nature.

I had dinner one night with a diplomat from one of the world's developed nations who has been active in the international climate negotiations. He explained to me in great detail the progress of those difficult talks and expressed hope that they might produce a series of trade arrangements between the North and South, so that clean energy technologies could at least begin to be sold to the rapidly industrializing countries of Asia and Latin America.

The limited and conventional scope of his vision made me impatient. "That strikes me as way too inadequate a response," I said. "Your thinking is too small. Given the magnitude of the climate threat, the solution has to be of a much higher order of magnitude. We need a leap."

He stared at me for a long moment, then shrugged. "You're right," he said quietly. "Unfortunately, there is no way we negotiators can get from here to there. But yes, you are absolutely right. I agree with you. We need a leap."

One world-class scientist suggested that "Maybe the best argument involves the spread of small wars over resources. The U.S. is still the world's policeman—whether or not you think we should be. Maybe that's the tack to take in arguing for change," he said, then paused. "The major argument that needs to be made is in terms of morality. It may be cost-effective, for example, to relocate the inhabitants of the Marshall Islands and Bangladesh [from their flood-prone homes], but it wouldn't be moral. I would love to see the churches become involved in this issue. They should be involved in the struggle," he groped, "maybe that's where support will come from."

I had a similar conversation with a high official in a major center of finance who spoke encouragingly about the possibilities of the kinds of "joint implementations" being discussed at the international level.

When I told him I did not believe those kinds of measures were enough to stave of significant climate disruptions, he reluctantly nodded his assent. "You have to realize there is a lot of institutional resistance to overcome. There are cavalier attitudes and obsolete thinking in the fossil fuel industry. But," he looked at me, his eyes suddenly less tired, "there will be progress and it will come from people fighting for change. That's how paradigm shifts take place."

But how do we achieve change on such a scale?

To reduce greenhouse emissions by 60 percent will take, in whatever form, a massive, sweeping change whose proportions are as monumental as the threat itself. And it must be driven by an urgency that recognizes that planetary systems change not gradually but with abrupt and devastating shifts.

What will it take?

In an ideal world we would have a ten-year plan to phase out all fossil fuel burning. We would have a brain trust—of industrial leaders, engineers, government officials, energy specialists, and parents—who would evaluate the various alternative energy forms. They would decide which kinds of renewable, climate-friendly energy are appropriate for different uses and different settings. Business planners and engineers would determine how best to produce and install them. And the public relations industry would put its extraordinary expertise to positive use to promote the acceptance of renewable energy around the world.

Using the same workforce that currently runs the coal and oil companies, an international enforcement agency of governments would assume control of the annual trillion-dollar-plus revenue stream from oil and coal. That revenue would gradually be funneled into an international fund to finance the production of renewable energy: windmill factories in India, fuel cells manufacturing in Germany, vast solar farms in the Middle East, hydropower turbine production in Thailand, and hydrogen gas plants in Cleveland.

Even in an ideal world, financing such an energy transition would succeed only if coal and oil revenues would be directly transferred to all the countries of the world to develop and manufacture and deploy those climate-friendly power sources that best use their own resources and best meet their own needs. Unfortunately, if the transfer of climate-friendly energy technologies is structured as yet another set of economically market-based transactions, it will fail—and our viability as an organized civilization will fail with it.

What would it take to accomplish such a transition and still prevent us from a freefall into economic chaos? It could be that an approach involving a few-cent tax on fossil fuels could be imposed to repay, at an equitable market price, all the people who own stock in coal and oil companies, as well as the lenders who have underwritten new power plants that would be converted, before the expiration of their lifetimes, to renewable energy facilities.

The flow of revenues resulting from the worldwide commerce in coal and oil would be more than enough to jump-start renewable energy manufacturing plants around the world. It would be enough to fund the teaching of millions of people in how to build and distribute and install the systems. There might even be enough to provide aid to those countries and industries that are most dependent on the fossil fuel trade and would be most heavily affected by its phase-out.

Remember the dimensions of the threat. The time limit is crucial. Huge corporations do not willingly surrender control. All our institutions—both private and government—have become accustomed in this fast-forward age to responding to problems only when they become emergencies. But by the time the impacts of climate change become more immediate and urgent, it will be far too late to affect them in meaningful ways. Carbon dioxide remains in the atmosphere between one and two centuries. It will continue to disturb the climate long after we stop burning oil and coal.

Still, there are precedents for governments joining together to regulate an industry. The Montreal Protocol was undertaken to phase out completely the manufacture and sale of ozone-destroying chemicals.

The Montreal Protocol, of course, was of a much smaller scale. The refrigerant chemicals in question were far less essential to our existence than oil and coal. Their manufacturers could afford to collaborate in their phasing out without sacrificing their relative competitive standings.

The ideologically conservative supporters of big coal and big oil repeatedly express dire fears of world government. With all due respect to their concerns, even this seemingly outlandish approach does not involve world government. We do not need to cede our sovereignty. International governance is far different—and far more limited. It is the control by a panel of governments over one discrete area of commerce whose destructive consequences are global in scope and transcend the reach of any individual national government. We didn't surrender our sovereignty to Canada or Belgium or Morocco when we agreed to an international ban on ozone-destroying compounds in Montreal. The United States is just as sovereign and independent today as it was before it signed the Protocol.

Such an approach might, in fact, enhance both our national security and our personal economic freedom. Most renewable energy systems are decentralized. The fuel cell in your basement unhooks you from the electricity grid. In decentralizing our energy systems, we remove a glaringly vulnerable target of terrorist sabotage. They could no longer cripple the East Coast, for instance, by disabling an unprotected system of power lines. In a stroke we would eliminate from our political landscape a huge, centralized, monopolistic bureaucracy that extends from oil fields and coal mines to our local utilities. A substantial piece of our personal economic security depends today on its business decisions in which we have no voice. In determining how to meet our individual energy needs, we gain both economic freedom and personal control.

One exciting thing about this Manhattan Project idea is that a ten-year effort to rewire the world would create millions and millions of jobs in every country on earth. The labor force required to outfit every home and building with solar energy panels or fuel cells or hydrogen fuel would be bigger than armies. Huge numbers of men and women could be easily taught to manufacture and install a whole variety of efficiency technologies and energy devices.

And renewable energy could well succeed high technology as the central driving engine of the global economy.

According to some calculations, for every million dollars spent on oil and gas exploration, only 1.5 jobs are created; for every million on coal mining, 4.4 jobs. But for every million spent on making and installing solar water heaters, 14 jobs are created. For manufacturing solar electricity panels, 17 jobs. For electricity from biomass and waste, 23 jobs.

The Ecuadorans outside of Guayaquil are not unique. Hundreds of millions of people languish every day, starving for work. Work is what will restore to them a measure of control over their lives. It will reduce their family size by providing a minimum level of economic security. Work is what their beaten dignity demands to be restored.

If the United States, in some miraculous way, could commit to a global energy public works project, it would jump-start economies all over Latin America, Africa, and Asia. Not only that, it would generate a profusion of new jobs for those Americans whom economists call marginal—people who, in other words, personify our national shame. It would bring them at least into the same world as those of us who can afford to buy a book.

The promise of a boom in energy employment may not be relevant to the managers and executives and professionals of the middle class. But many of them do lack a sense of purpose in their lives, an element of meaning in their work beyond the thrill of the hunt for the next corporate deal. Many of them yearn for a meaningful endeavor that extends to the larger world. An outpouring of social and intellectual energy could very well result if we mobilize our human and financial resources around the energy transition. To jointly help mend the planet could, I think, bring a unifying mutual recognition to a fractious and divided society. It could involve us together in a bond of common responsibility.

The global climate crisis draws together three dimensions of enormous scope and pervasive impact. First, its natural dimension is of truly cosmic proportions. Second, its central economic dimension is the widening global fault line between rich and poor that threatens to split humanity irreparably. Third, the energy dimension is so central to society as to constitute its nervous system. Energy heats our homes, powers our vehicles, electrifies our offices and schools, and fuels our manufacturing. As the central element of our global infrastructure, it shapes basic ways in which we live.

These three fascinating and deeply engaging dimensions challenge both our habits and our intellects. Forging a real solution could push us to grow enormously in very many ways.

Truly addressing the climate problem must involve us all in redesigning our economy to conform to planetary limits. It would call upon us to account for the depletion of resources and for damage to the natural systems using hard numbers. Our own collective growth—and its environmental consequences—have left us no other choice.

But we cannot redesign the global economy without addressing the nonenvironmental issues that affect the economic and political conditions of society. Underlying any alteration of our economy is the need to assign proper roles to competition and cooperation. The challenge here is to strike the optimal balance between competition and cooperation, so that new heights of human achievement may be generated while the baseline conditions for human peace are extended to the entire world.

I don't know what institutions that balanced economy would encompass. We would need a global dialogue to figure it out. But a wealth of helpful economic and social science literature is available, many superb minds are around to tap, and many concerned individuals are eager to help. Meeting the challenge would reward us with a sense of meaning and excitement and accomplishment in our public lives.

It would require a jolt of stand-up social courage from us all. For one thing, this undertaking will have costs. The most obvious costs will be felt by the nations and companies that most depend on fossil fuels for their economic wealth. Australia would lose a lot of money from the coal it exports. Saudi Arabia, Kuwait, and the other oil-producing nations would become poorer relative to other countries. They could recover some measure of their loss by constructing vast solar farms. But their oil wealth may well prove to have been a fragile historical bubble, a fifty-year windfall for the Middle East. (The plan might also cause unease in Israel. As the strategic importance of Middle Eastern oil declines, the U.S. commitment to that country might decline as well.)

Ancillary businesses—oil rig construction firms, exploration crews, gasoline additive companies—all would suffer. Perhaps the massive flow of public revenues from oil and coal commerce would allow for funds to offset those costs. Those kinds of calculations require far more knowledgeable minds than mine. In any case, the nations and companies most impacted by the switch would need some sort of aid—but without compromising reaching the target of zero emissions in ten years.

The energy transition should not threaten the livelihoods of the thousands of people who currently work in the coal and oil industries. They would continue to be employed, keeping the diminishing flow of fossil fuels coming as alternatives come on line and are expanded. When, say, in the second half of the target decade, the oil and coal industries began to shrink, their workers could be retrained and given jobs in the new renewable energy industry.

This plan is simple in its elements, but it is huge in the institutional changes it implies. It alters the balance between governments and corporations and requires unyielding public regulation to be imposed on a specific segment of the free market. And while it may be wrong in its approach or in its details, all my reporting on climate change convinces me it is at least of the right order of magnitude. The far more limited and incremental measures now being discussed by diplomats and corporate leaders are, I believe, based on the narrow business-as-usual perspective. They fall far short of what is needed to stop the melting of the glaciers and stabilize the ice shelves of Antarctica.

The conventional political wisdom of the late twentieth century places precious little faith in people's will and courage to leap beyond the boundaries of their own selfishness. I don't accept that view. When people first learn about the warming of the planet, most of them react with fear but quickly rebound with a strong desire to act. What stops them is not a lack of will or concern, but rather the absence of a channel for their energy.

The first and clearest way to provide them with a channel is to put the climate crisis—in all its massive dimensions—at the center of the public stage. Let the media cover the scientific evidence of climate change the way they now cover political campaigns. As scientists gain new understandings of the rates and specific impacts of climate change, they should be featured prominently in the news media. So should new signals of stress from the forests and the oceans and the plains. Let academic consortia and business roundtables and congressional committees and television talk shows put the precarious condition of our beleaguered planet in the spotlight of public attention.

Perhaps people are so fragmented, so alienated from one another, that they would shrink from the challenge and block out the message. Such self-imposed ignorance could speed our descent into a more brutish and storm-battered existence. But I don't think they would react that way. I think if most people were exposed to the truth and given a channel through which to act, they would choose responsibility over selfishness. I think if they truly understood what is happening to the earth, there would be an outpouring of will and courage.

What I want most is the opportunity to find out.

To a small extent, the energy transition will challenge engineers and industrialists and planners. They will be in the forefront of selecting the mix of new energy technologies that must be on line a decade from now. They must determine how best to jump-start their production and plan their installation in every building and vehicle in the world.

To a greater extent, it will challenge economists, intellectuals, activists, and political leaders. They must guide to fruitful conclusion the inclusive discussion of how to reengineer the global economy so it can coexist with the global environment—so that both can heal and thrive.

But the greatest challenge of the energy transition, I think, will be to the collective self-image and sense of responsibility of all of us.

It is easy to see how much damage human beings can do—we need only survey the condition of planet. But it is harder to see how much good we can do. For that, we have to look inward, to the dictates of our morality.

We are holding in our hands not only the health of our endangered planet but our own future as a civilized species. History highlights our limitless capacity for cruelty and devastation and mutual destruction. People have hated each other for many reasons, and they have expressed that hatred in many ingenious ways. Our awareness of this history should highlight, with equal brilliance, our powers to nurture and heal and create.

We are no longer dependent children of the earth. Nor are we, any longer, its adolescents, dangerously out of control. The message of the oceans is that we have become adults now, parents of the future and caretakers of the planet. We are standing at the confluence of massive forces of human and planetary history. We have become as powerful as any cosmic force on the planet—the proof is as vast as the sky we all share. Having grown up as a species, it is time to grow up as individuals. It is time to grow up and step out, for there is a great job to be done.

On my back porch, on the last day of June 1996, it is chillier than normal. From the dense thicket behind my backyard, the slender overarching trees soar toward the westward-tilting sun, their carbon-flushed leaves brilliantly emerald in the afternoon light. The headline stripped across the top of today's *New York Times* declares that this year's wildfire damage in the United States is three times greater than normal. With severe drought gripping Arizona and New Mexico and portions of southern California, Colorado, Nevada, Texas, and Utah, 68,000 wildfires have burned. The normal annual average is 40,000 wildfires. In just the first half of this year, 2.3 million acres were lost to fire.

A second story, this one on New Orleans, updates a report that appeared a year earlier in *The Boston Globe.* It contains a few paragraphs worth quoting:

> *"On even the calmest days, a limb from a 200-year-old oak will just snap and land with a crash on the street. On scenic St. Charles Avenue, where the old green streetcar rattles between rows of ancient oaks, whole trees have just toppled over, dead. The odd thing is that "they look perfectly healthy," said Flo Schornstein, director of the New Orleans Department of Parks and Parkways. "But they're hollow on the inside."*
>
> *The trees, after so many years without a normal killing frost, are haunted by bugs: millions and millions of tiny, white, blind bugs. The Formosan termite . . . has slowly and steadily infested at least one in every seven trees, and some experts believe the number may be closer to one in five. While city officials and entomologists are scrambling to find a way to stop the termite infestation, for now the insects are winning, and spreading by the millions.*

We are standing on the cusp of a very precarious future, as immediate as the lifetime of our children. There are those who want to swap it for their short-term gain, and they will if we let them. But we know too much to lay the blame on them. If we passively yield control of our collective destiny to those destroyers, we will all share the responsibility for the unspeakable consequences. If each of us chooses inaction, it will make us all accomplices in the triumph of greed and short-sightedness and selfishness.

We can no longer pretend to ignorance or innocence.

The future will be what we make of it now—or what we let it make of us all in a very short time.

APPENDIX

A Scientific Critique of the Greenhouse Skeptics

*MOST OF THE SO-CALLED GREENHOUSE SKEPTICS HAVE RE-*peatedly criticized the findings of the 2,500-member Intergovernmental Panel on Climate Change (ICPP) that indicate that planetary warming is occurring, that it is separate from the natural variability of the climate, and that it is caused in essence by the buildup of greenhouse gases—notably emissions from the burning of coal and oil.

Earlier chapters have documented the hitherto undisclosed funding that several of these self-proclaimed skeptics have received from fossil fuel interests, as well as their participation in industry-funded public relations activities. Still, private funding—even of an undisclosed nature—does not necessarily result in flawed science. Moreover, it is beyond the scope of a journalist to evaluate the findings of scientists.

In various forums, however, several well-established and highly regarded climate scientists have provided rebuttals to the greenhouse skeptics and critiques of their work.

This appendix contains responses to the scientific assertions of some of the most visible and prominent skeptics—Dr. Patrick Michaels, Dr. Robert Balling, and Dr. S. Fred Singer—by several leading climate scientists, including:

Dr. Michael MacCracken, former climate modeler at Livermore and currently director of the Interagency Office of the U.S. Global Change Research Program

Dr. Jerry Mahlman, director of NOAA's Geophysical Fluid Dynamics Laboratory at Princeton University and chair of NASA's Mission to Planet Earth Scientific Advisory Committee

Dr. Benjamin D. Santer, of the Program for Climate Model Diagnosis and Intercomparison at the Lawrence Livermore National Laboratory, and convening lead author of chapter 8 of the 1995 Working Group I IPCC report

Dr. Tom M. L. Wigley, senior scientist at the National Center for Atmospheric Research

CONGRESSIONAL TESTIMONY OF MICHAEL MACCRACKEN

On March 6, 1996, Michael MacCracken submitted prepared testimony to the Committee on Science of the House of Representatives. One part of that testimony addressed recurring criticisms by the skeptic scientists of IPCC findings that corroborate increased atmospheric warming and attribute that increase to human emissions of greenhouse gases. What follows are his remarks.*

Comments on a Selection of Scientific Statements about Global Climate Change

In the recent IPCC assessments, an interesting revelation has been the difference in confidence levels that are expected and are justified

*U.S. House of Representatives Committee on Science, *Hearing on U.S. Global Change Research Priorities: Data Collection and Scientific Priorities,* March 6, 1996.

by the present state of knowledge. For IPCC WGI on the expected changes in climate, traditional scientific analyses are seeking high levels of certainty (e.g., 90 to 95 percent) before drawing conclusions. . . .

For WGII on the impacts of climate change, levels of certainty tend to be lower, but still significant. . . .

For WGIII on economic implications of change, only rough central estimates that are acknowledged to be incomplete are available.

What is most surprising about some of the statements [of the greenhouse skeptics] is that they tend to focus on the results that are most certain in the IPCC findings, findings for which mainstream scientists are arguing about whether the odds of their central predictions are 5 to 1, 10 to 1 or even 20 to 1. While by no means all aspects of the climatic effects can be estimated with such certainty, such levels are well above those for which many societal decisions are made. . . .

There will always be critics and those not fully satisfied, and indeed there are both those who believe the conclusions of the IPCC should be weaker and those who believe the conclusions should be stronger. However, at this time, the IPCC results are the most thoroughly reviewed and considered views on global climate change and they deserve very high respect.

The very great importance of this issue for society in terms of the threat of environmental change and the potential for changing aspects of important societal activities has generated considerable public discussion of both the findings and the uncertainties concerning climate change. Trying to iron out dueling perspectives through the media and otherwise outside the scientific process will never be satisfactory.

[In response to assertions by skeptics that global warming projections are contradicted by temperature records:]

Data for the full year clearly indicate that the global average surface temperature for 1995 was near or actually the warmest on record—and was likely to be warmest since 1400 and even much earlier. What is most interesting is that 1995 set (or matched) the recent warming record even though there was no El Niño, the solar cycle was near a minimum and ozone depletion was near record levels, whereas in 1990 (the next warmest year) these comparative cooling influences were not present.

[In response to assertions by skeptics that the climate models are inaccurate and are contradicted by temperature data:]

Global average temperatures have risen about .5 degrees C (about 1 degree F) during the past 140 years. . . . As clearly summarized in the IPCC Second Assessment, the results of model simulations that include both greenhouse gases and sulfate aerosols, to the extent that we understand them, show good agreement with the observed record.

[In response to skeptics' claim that the most recent IPCC projections, which include a lower boundary in the range of expected warming, reflect inadequate methodologies:]

The IPCC lower bound estimate of warming for the year 2100 has come down about 0.7 degrees C since their 1992 estimate—not due to model uncertainties or improvements, but to changes in the presumed scope of human activities. The slightly lower estimates of warming, which are within the earlier range of uncertainties, had much less to do with uncertainties in the climate models or their responsiveness to human activities, than with the estimates of social activities. To a first order approximation, the differences were caused by the inclusion of sulfate aerosols and the early phasing out of CFCs in order to protect the ozone layer.

[In response to skeptics' claim that satellite measurements contradict the rates of warming predicted by computer models:]

Regarding the supposed difference in the [satellite] and model-predicted rates of warming, such comparisons that have been done have been of fundamentally different quantities. To make the comparison, the model run that is used has been adjusted in several ways that are inappropriate. Quite simply, the comparison is totally inappropriate to do and is presented in a misleading way.

[In response to skeptics' assertions that there is no firm signal of global warming due to human activities:]

Human-induced climate change is now "discernible." A major scientific advance in the last few years has been the ability to search for patterns of climate change, not just changes in the global average temperature. This has been possible because more realistic cases have been run using the climate models. The IPCC conclusion recognized the following five points:

(1) the global average temperature of the latter part of this century is at least as high as for any time in the last six hundred years (the period for which we can reasonably construct temperature changes);

(2) the temperature trend has been upward since the last century;

(3) the latitude-longitude pattern of [the] global temperature trend is unlike patterns associated with natural variability (including solar variations) and quite similar to the predicted response in models to increasing carbon dioxide and sulfate aerosol concentrations;

(4) the vertical and pole-to-pole pattern of temperature change is unlike that of natural variability (including solar and volcanic variations) and is in general agreement with changes from greenhouse gases, sulfate aerosols, and ozone depletion; and,

(5) the magnitude of the changes is about as estimated by the models.

While no one or two of these would be as convincing, the IPCC concluded, rather conservatively, that the "balance of evidence suggests that there is a discernible human influence on global climate."

TOM WIGLEY'S REPLY TO PAT MICHAELS'S CONGRESSIONAL TESTIMONY

At a hearing before the Subcommittee on Energy and Environment of the House Science Committee on November 16, 1995, Patrick Michaels—one of the most visible and prominent of the greenhouse skeptics—provided extensive testimony purporting to show that the findings of the IPCC were scientifically flawed. Tom Wigley has written this critique of that testimony.*

On November 16, 1995, Patrick J. Michaels, an associate professor in the department of environmental sciences at the University of Virginia, testified before the Subcommittee on Energy and the Environment, U.S. House of Representatives, on issues related to human-induced (or anthropogenic) climate change. The following is a critique of the scientific arguments presented in that testimony.

Since the testimony deals with a scientific issue, I have treated it as I would any scientific document or manuscript. For readers unfamiliar with the process through which scientific manuscripts come to be published, the normal procedure is for such manuscripts first to undergo peer review—i.e., they are first judged (or refereed) by other specialists in the field for the soundness of their scientific content. Usually issues of content and presentation are raised in this refereeing step; these are then addressed and dealt with by the author, and if the revisions are satisfactory, the manuscript is published. Manuscripts may have to go through this process a number of times before reaching a state suitable for publication. Not infrequently, the original criticisms may be so severe or the responses to them so poor that the manuscript may be irreparable. Under such circumstances the manuscript will be rejected by the chosen publication medium.

The testimony of Michaels has not been peer reviewed; such testimonies never are and are not required to be. However, since the main issues he addresses are scientific issues, it is, in this case, quite appropriate to apply normal scientific refereeing procedures to assess the scientific credibility of his testimony.

CRITIQUE

MICHAELS: "My testimony centers around four issues of critical importance to the problem of global warming.

*Provided to the author by Tom M. L. Wigley, August 1996.

1. "New calculations support the view of the scientists who predicted that global warming would be relatively modest."

The truth or otherwise of this statement depends on: what "new calculations" are being referred to, who the "scientists" referred to are, what "relatively modest" means, and whether or not Michaels is correctly interpreting these "new calculations." As shown below, his interpretation is seriously flawed.

From the remainder of Michaels's testimony, it is clear that "the scientists" are those who "for the last decade" have "argued based on observed data . . . that the modeled warming was too large." This group includes Michaels and other so-called greenhouse skeptics like Robert Balling and Sherwood Idso.

Just what does "too large" mean? Michaels goes on to claim (incorrectly) that the majority of scientists (i.e., persons other than Michaels, Balling, and the like) think that the equilibrium warming for a doubling of the CO_2 concentration is 4 degrees C, while the observational data suggest a warming of only 1.0–1.5 degrees C for $2 \times CO_2$. Presumably, it is the 4 degrees C that is "too large."

In fact, the CO_2-doubling temperature change, or climate sensitivity, is not 4 degrees C but has been thought, for many years, to be in the range of 1.5–4.5 degrees C. This range was endorsed by the IPCC in 1990 and subsequently. The IPCC gives 2.5 degrees C as the best estimate, not 4 degrees C. Michaels is either ignorant of this or he deliberately chooses to ignore it, because he later states: "A warming of 2.5 degrees C for doubled carbon dioxide" is "a relatively low figure." But 2.5 degrees C is not "relatively low." It is the current best estimate!

Michaels's claim that the observational data suggest a climate sensitivity value of 1.0–1.5 degrees C is a seriously out-of-date estimate. Furthermore, he implies, by quoting this rather narrow range, that one can deduce a quite accurate value of the sensitivity from the observational data. If only this were so! Unfortunately it is not, and there are very large uncertainties in trying to back out a sensitivity estimate from the data.

In 1990, the IPCC (Wigley and Barnett 1990) noted that, if all the observed warming since the late nineteenth century were due to anthropogenic greenhouse-gas forcing (note the *if*!), then the sensitivity would be around 1.5 degrees C. Large uncertainties surround this estimate, due mainly to two factors: uncertainties in the magnitude of anthropogenic forcing; and uncertainties in the contribution of natural variability to the observed warming (see, e.g., Wigley and Raper 1991.) For example, if we have overestimated the forcing (see below), then a higher sensitivity would be implied.

As an aside, I note that Michaels refers to some of his own work (Michaels and Stooksbury 1992) that relates to the issue of determining the climate sensitivity empirically. The Michaels and Stooksbury paper does not appear to be his source for the sensitivity estimate he quotes—presumably he gets this out-of-date estimate from the 1990 IPCC report. Nevertheless, his paper does address the more general issue of assessing the consistency between observations and model simulations of past global-mean warming.

There are two approaches used for such consistency testing. The first is to run an appropriate model with the observed (albeit uncertain) forcing record and compare this with observations. The second is to run a number of model simulations with different climate sensitivities, compare the results of each run with the observations, and see what sensitivity gives the best match with observations. A good match in the first case, or a best-fit sensitivity in the range of 1.5–4.5 degrees C in the second case, essentially validates the result estimated independently from sophisticated climate models.

Michaels's work uses the first approach. He does not use a model in the conventional way, however, but tries to make use of the previously published work of Manabe et al. (1991). Manabe's work is not an analysis of past warming, but an entirely different type of experiment; so Michaels has to modify the Manabe results in some way to twist them into an appropriate form. What he does is stretch out the Manabe et al. simulation results in an attempt to adjust the effective forcing down to that which has been observed to date.

He does this by considering only the initial part of the experiment, over which the total forcing is similar to what Michaels assumes to be the observed forcing. To illustrate the principle, in the Manabe experiment the rate of change of forcing is around 0.6W/m^2 per decade. The observed forcing due to greenhouse gases alone (i.e., ignoring aerosol effects) is about 2.6W/m^2 since preindustrial times, of which 2.1W/m^2 has occurred over the past 100 years. If we consider only this latter period, the rate of change of forcing is about 0.2W/m^2 per decade. Thus, following Michaels's logic, if one only considers the first 30 or so years of the Manabe experiment, this should simulate what has occurred in the real world over the past 100 years. In other words, a real-world analog could be obtained by stretching the time scale of the Manabe experiment out by a factor of around three.

It is not possible to go into the technical details here, but this stretching-out method is seriously flawed: it is simply not possible to translate results from one forcing experiment to another in such a simplistic way. Here are a few of the problems.

First, the response of the climate system to any externally imposed forcing lags considerably behind the equilibrium (i.e., eventual) response, due to the thermal inertia of the oceans—in the same way that a car does not immediately jump to its top speed when the accelerator pedal is floored, due to the mass inertia of the car. This lag is greater the faster the rate of change of forcing, and cannot be translated from one forcing case (such as the 0.6W/m^2 per decade in Manabe's experiment) to another (e.g., the 0.2W/m^2 in the real world). Second, in the Manabe case, with its relatively large lag and slow response relative to equilibrium, the contribution of the model's own internal variability (or noise) is large relative to the slowly evolving signal for at least the first decades of the experiment. What Michaels stretches out, therefore, is not a pure response signal, but a signal that is highly contaminated with the natural variability noise that is inherent in the Manabe model. Third, not only does the response depend on how rapidly the imposed external forcing changes, but it also depends in a very complex way on the precise history of forcing. In the Manabe experiment, the forcing change is a linear one. In the real world it is far from linear—sufficiently so that the linear case is a very poor analog for what has really occurred. Fourth, Michaels uses an incorrect history of past forcing. By considering only the greenhouse-gas component, he grossly overestimates the forcing that has actually occurred. Fifth, the model he bases his stretching on has a sensitivity well above the current best estimate.

In summary, Michaels's stretching method is simply not a credible way to interpret the Manabe model results. There are, furthermore, well-known and well-used methods available that he could have used; methods that have been used by many others (including IPCC) for examining the issue of the consistency between model predictions of past warming and what has been observed. The IPCC conclusion, based on these methods, is that there is no inconsistency—see, for example, Figure 8.4 in the 1995 IPCC Second Assessment Report (Santer, Wigley, Barnett, and Anyamba 1996).

Michaels goes on to note that his work was not taken into account by the 1992 Rio Earth Summit meeting, even though it was published before the ratification of the Framework Convention. While not taken into account, his work has been considered, and judged to be irrelevant. His work simply does not pass muster scientifically; so it is not surprising that it was ignored in preference to the comprehensive review documents produced by IPCC.

The "new calculations" Michaels refers to are experiments performed by the U.K. Hadley Centre (Mitchell et al. 1995) that include the negative forcing effect of sulfate aerosols, which act to offset part of the anthropogenic greenhouse forcing. These model calculations show a twentieth-century warming that is similar to that observed (see Kattenberg et al. 1996, Fig. 6.3). Since the Hadley Centre model has a sensitivity of 2.5 degrees C, the similarity between the modeled results and observations implies that the real-world sensitivity is similar to the model sensitivity.

While these *are* new experiments, the results are not at all new. The fact that including aerosol forcing effects leads to a larger sensitivity than would be implied by greenhouse-gas forcing alone was first noted (and quantified) in 1989 (Wigley 1989). The observationally based sensitivity estimate was first refined from the 1990 IPCC estimate to account for aerosol effects in 1992, by Wigley and Raper (1992), who gave a central estimate of 3.4 degrees C.

It is clear that including aerosol effects changes the implied climate sensitivity considerably. Since the magnitude of the aerosol forcing is highly uncertain, it is not surprising that the empirical, observationally based sensitivity, which is extremely sensitive to the assumed aerosol forcing magnitude, also is highly uncertain. Because of this uncertainty, the results of Mitchell et al. (1995) and Wigley and Raper (1992) are entirely consistent—quantitative differences between them reflect the above-mentioned uncertainties. Indeed, using the *latest* estimate of the aerosol forcing term, results given by IPCC imply a sensitivity well above 4.5 degrees C if all of the observed warming is assumed to be anthropogenic (Santer, Wigley, Barnett, and Anyamba 1996, Fig. 8.4).

Contrary to Michaels's claim that the data imply that models have overestimated the warming, the opposite may well be the case! It would, however, be either misleading or foolish to assert that the observational data allow us to derive a better estimate of the sensitivity than that given by models. When uncertainties are properly accounted for, the observationally based sensitivity could be anywhere between about 1.5 degrees C and considerably more than 4.5 degrees C. Basically, there is no inconsistency whatsoever between observations and theory. In fact, the consistency is quite remarkable. Santer, Wigley, Barnett, and Anyamba (1996, Fig. 8.4) compare the observed warming with modeled warming using current best estimates of anthropogenic (greenhouse gas plus aerosol) and natural solar forcing. For a climate sensitivity of 2.5 degrees C, the agreement is strikingly good.

So where has Michaels gone wrong? All of the information given above was available to him at the time of his testimony, so having incomplete information cannot be the explanation.

His claim that the consensus sensitivity is 4 degrees C probably comes from thinking that the Manabe et al. (1991) model, which has a sensitivity of about 4 degrees C, is somehow typical. Using a single model result rather than the comprehensive review result from the IPCC is somewhat perverse, and certainly misleading.

Why does he think that future warming may be "relatively modest" and that "lower numbers [appear] more likely to be correct"? He even states, entirely erroneously, that this view is "acknowledged [by] the community," and that the 1995 IPCC Second Assessment Report "now states that the climate models" used in 1992 "were in fact predicting too much warming." Michaels's misconceptions here reflect some serious misunderstandings of the nature of climate modeling experiments.

First, Michaels mixes up models and experiments. Referring to the Mitchell et al. (1995) work, he says that these authors "ran two types of climate models." In fact, what they did was run two types of experiment with the *same* model!

The way climate modeling works is like this: For any given model (the current Hadley Centre model is just one example), the results it produces depend on how it is forced. Basically, the response of any system (real or modeled) depends on how hard one hits it; so one can carry out a series of experiments hitting (or forcing) the system (i.e., the same model) in different ways.

The response of a model to a particular forcing also depends on how sensitive the model is to external forcing; i.e., on its climate sensitivity. Different models have different sensitivities (and most lie in the range of 1.5–4.5 degrees C equilibrium warming for the forcing that corresponds to a CO_2 doubling).

It is clearly important not to confuse (as Michaels does) the *response* in a particular case with the model's *sensitivity*. The current Hadley Centre model has a sensitivity of around 2.5 degrees C. When forced with the past history of greenhouse-gas forcing, it simulates a warming that is greater than that observed. When forced with greenhouse-gas-plus-aerosol forcing, it simulates a warming very similar to that observed. Michaels refers to these two *experiments* as different *models*. Further, he goes on to imply that these two "models" have different sensitivities. In fact, there is just one model; and it has a single sensitivity (of around 2.5 degrees C, as noted above). When the model is forced with a lower forcing (i.e., when aerosols are included), it simply and quite naturally gives a smaller warming response.

Michaels compounds this misrepresentation by implying that one can estimate the climate sensitivity (which is an equilibrium concept) directly from the time-dependent (or transient) model simulations of the Hadley Centre. This is incorrect. The reason why the sensitivity cannot easily be inferred from the transient run of a model like the Hadley Centre model is because the model's transient response lags behind its instantaneous equilibrium response, due to the thermal inertia of the oceans; in particular, when the model forcing reaches that for a CO_2-doubling, its response will lie substantially below the equilibrium response for that CO_2 level (i.e., considerably below the model's climate sensitivity). Thus, estimating the model's sensitivity from a transient experiment is a nontrivial task. One cannot glibly pluck a number out of the air, as Michaels does.

To compound these errors, Michaels also seems to confuse the effect that including aerosols has on warming (i.e., it leads to reduced warming) with the implications of such an effect for the climate sensitivity (i.e., it implies a larger sensitivity). Together, through these two factors, aerosols have brought models and observations into much closer accord than they were previously. They have not made us rethink our previous results or adjust our estimates of the climate sensitivity. What they have done is just the opposite—they have improved our confidence in the standard sensitivity range (1.5–4.5 degrees C) and, hence, in model projections of future climate change.

But all this is, in fact, irrelevant. The issues of individual model sensitivities, and empirical estimates of the sensitivity, have absolutely no bearing on the future global-mean-temperature projections published by the IPCC, since such model projections have taken no account of empirical sensitivity evidence (due to its large uncertainty). Instead, projections are given for a wide range of possible sensitivities. Thus, the new evidence that Michaels discusses has had no influence at all on such projections—beyond increasing our confidence in them.

I will now return to the original statement No. 1 by Michaels. Do the latest projections suggest only a "modest warming"? The answer is categorically no. It is true that the latest projections (Kattenberg et al. 1996) are less than those given by the IPCC in 1990—but not for the reasons Michaels indicates.

The latest projections available at the time of Michaels's testimony (Kattenberg et al. 1996) are for a global-mean warming over 1990–2100 of around 2 degrees C with an extreme range of 0.8 degrees C–4.5 degrees C. While it is true that these results are slightly smaller than the projections given by the IPCC in 1990, the important thing to note is that they are not directly comparable with these earlier results. This is because the 1990 results were based on different emissions scenarios, scenarios that differ markedly from the 1992 emissions scenarios. The 1992 emissions scenarios, furthermore, now include SO_2 emissions, which lead to the production of sulfate aerosols with, in most cases, an attendant cooling effect (albeit relatively small).

What does a warming of 0.8–4.5 degrees C mean? Can it in any way be considered modest? At the low end of the current range, the warming could, perhaps, be considered modest; but I note that it still implies a warming rate almost double the warming rate that has prevailed over the last century, and that the probability of such a small warming is *very* low. At the high end of the range, the warming would be considered by most to be extremely large. In the middle of the range, a 2-degrees-C warming by 2100 represents a warming rate some *four times* that over the last century!

Why does Michaels stress the Hadley Centre model results? He implies that future climate projections given by the IPCC are either based on this model, or dependent on its veracity—but this is wrong. IPCC warming projections have never been based on nor dependent on a single General Circulation Model with a single sensitivity. To the contrary, all global-mean-temperature projections given by the IPCC have included estimates of uncertainties arising both from emissions uncertainties and from model-based uncertainties in the climate sensitivity. Complex models like the Hadley Centre model cannot explore such uncertainties, so the IPCC has made use of simpler models for this purpose. Specifically, the model used is that of Wigley and Raper (1987, 1992), which has been tested against a number of more complex models. Just because the Hadley Centre model and a number of similar models have deficiencies (which are well known, and extensively discussed by the IPCC) is irrelevant: uncertainties arising from such deficiencies are fully accounted for in IPCC calculations.

Another misconception that Michaels propagates is the idea that some radical change has been made in the performance of the Hadley Centre model between 1990 and now. This, too, is wrong. The model has been changed, but not in any way that is relevant to the debate. The most important change that has occurred has been not to the model but to how the model is forced (viz., now including aerosol effects). As for the model itself, the main change has been in its coupling to a realistic ocean model so that time-dependent simulations can be performed.

Michaels also claims that the earlier version of the Hadley Centre model was used to make estimates of present temperature, but this is incorrect. The earlier model could not be used for this purpose—to do so requires coupling the model to a full ocean general circulation model; and only the latest version of the model has this capability. Furthermore, because the earlier model and the more recent model have similar climate sensitivities, if the former *could* be used to estimate present temperature, it would give similar results to the latest model.

Michaels's statements on this subject are a catalog of misrepresentation and misinterpretation. To conclude from his arguments, as he does, that "the scientific review process that [forms the basis of] international agreements has been flawed" is completely unjustified.

2. "Older calculations that were based on the 1992 Framework Convention on Climate Change were known to be greatly overestimating warming at the time the convention was ratified."

This is entirely untrue. To see this, one must compare the 1990, 1992, and 1995 global-mean-temperature projections. Michaels's criticism relates to the 1992 versus 1995 comparison, but the earlier results are still of interest. It is true that calculations published in the 1990 IPCC report gave larger warming amounts than the latest values, but there are no qualitative differences. The differences do not arise from any change in models or in our understanding of the climate sensitivity; rather they arise from the use of different forcing scenarios for the future. The earlier projections were based on different emissions scenarios. They are simply not directly comparable with the latest results because of this and because aerosol effects were ignored.

As regards the 1992/1995 comparison, calculations using the *same* (IS92) emissions scenarios as used in the latest work *were* included in the 1992 IPCC report (pp. 171–75), but even these two sets of results are not directly comparable because the 1992 results did not include aerosol effects. Even so, for changes over 1990–2100, the most recent values are only from 0.2 degrees C to 1.0 degrees C (13 percent to 29 percent) below the earlier values (cf., Fig. 6.22 in IPCC95 with Fig. Ax.3 in IPCC92). If aerosols had been included in the 1992 calculations, then the 1992 and 1995 results would be even closer.

It is entirely false to say that the earlier results, in Michaels's words, "greatly overestimate[d]" the projected warming. There is no qualitative difference between any of these sets of results (i.e., the 1990, 1992, and 1995 IPCC results). They all imply a very large warming relative to natural variability and past warming. It is also false to claim that the IPCC knowingly published overestimated values. The 1992 report clearly states (p. 193) that the results presented there are given specifically to compare warming amounts for the 1990 and 1992 emissions scenarios using a common set of assumptions. The 1992 report was fully cognizant of aerosol influences. However, its main focus was not on climate change but on the driver for climate change, radiative forcing. Apart from a brief comparison related solely to the effect of the new (1992) emissions scenarios, the 1992 report *does not give* any formal revised projections of global-mean warming. Such revisions were not made until the 1995 report.

3. "Critical scientists are still being denied data (by taxpayer-supported organizations) that are required to quantitatively review new syntheses on global warming."

This statement contains two untruths. First, it is entirely unnecessary to have original "raw" data in order to review a scientific document. I know of no case at all in which such data were required by or provided to a referee. Thus, Michaels's claim that the "refusal to supply me the model results seriously compromises the scientific review process" is unjustified. No other reviewer (and there were over 100 for this part of the IPCC Second Assessment document!) requested such data. Second, while the data in question (model output from the U.K. Hadley Centre's climate model) were generated using taxpayer money, this was *U.K.* taxpayer money. *U.S.* scientists therefore have no a priori right to such data. Furthermore, these data belong to the individual scientists who produced them, not to the IPCC, and it is up to those scientists to decide who they give their data to.

4. "Therefore, any studies of the impact of climate change on ecosystems, health, and the economy, based upon older models, are in error, and newer models have yet to be properly reviewed."

The validity of this conclusion as a consequence of points 1–3 is seriously flawed, since points 1–3 themselves are flawed. Nevertheless, it is true at a pedantic level that "older models are in error"—models can never hope to be "correct," and their deficiencies are discussed at length in the various IPCC reports. But impact studies do not rest on the validity of climate models; most such studies are assessments of the sensitivities of various impact sectors to climate change in general and do not depend on the validity of any particular climate change projection. Furthermore, Michaels's claim that "newer models have yet to be properly reviewed" is incorrect. In fact, all "newer models" have been examined extensively (the process is called model validation), and their various strengths and weaknesses are both well known and discussed comprehensively in the IPCC reports.

In summary, many of the supposedly factual statements made in Michaels's testimony are either inaccurate or are seriously misleading. If I were to judge this as a scientific paper, or as a paper reviewing the science of climate change, I would have to recommend its rejection. The document contains so many errors, misconceptions, and misinterpretations that I can see no way that it could be revised to a state where publication might be considered.

CRITIQUE OF MICHAELS'S CONGRESSIONAL TESTIMONY BY JERRY MAHLMAN

In one area of his congressional testimony, Michaels criticized scientific model conclusions by comparing them to satellite findings on the climate record. Following that testimony, Jerry Mahlman addressed a number of Michaels's contentions.* What follows are Mahlman's responses to Michaels's assertions in a question-and-answer format with the congressional staff:

> Q: *During the course of the hearing, Dr. Michaels provided a chart that purported to show that a key model used by the IPCC did not provide adequate representation of temperature data in the 5,000–30,000-foot layer [of the atmosphere, data that had been] gathered by satellites over the past twenty years. Please provide your own views on the methodology and interpretation used in this chart and its significance . . . to the validity of the IPCC models.*
>
> MAHLMAN: Regarding Patrick Michaels's use of MSU [Microwave Sounding Unit] satellite temperature trends to evaluate global warming model calculations. First, it is clear that use of a fifteen-to-twenty-year record to determine temperature trends is an extremely dubious process. Even if all other things are equal, the natural variability of the climate makes it very difficult, if not impossible, to separate natural variability from systematic human-produced trends through the use of short climate records.
>
> All other things are not equal. The nearly twenty-year MSU temperature record contains a number of factors that were *purposefully* not considered in the GFDL [the NOAA Geophysical Fluid Dynamics Laboratory] climate model run that Michaels uses for comparison. Most importantly, the observed temperatures from 1976 to 1995 are affected

*Hearing Before the Subcommittee on Energy and Environment of the Committee on Science, U.S. House of Representatives, Nov. 16, 1995 (Washington: Government Printing Office, 1996).

by a number of complicating factors that include: possible problems with the MSU data set; shortness of the MSU temperature records; climate response to *previous* warming/cooling forcings; a cooling offset due to ozone losses in the stratosphere and upper troposphere; episodic cooling offsets due to massive eruptions of the El Chichon and Pinatubo volcanoes; and the still-uncertain cooling offset due to sulfate aerosol effects.

Finally, the GFDL climate model experiment that Michaels refers to was an IPCC-recommended idealized calculation of the response of a coupled model to a one-percent-per-year increase of CO_2 up to a doubling over pre-industrial levels. Such a model experiment is not designed to mimic a particular data set. Oddly, Michaels does not even set the zero line the same when he makes this inappropriate comparison in . . . his testimony. Even worse, it is unthinkable to do so for a twenty-year record when considerable natural variations of climate occur on ten-to-thirty-year time scales.

Q: *Is Dr. Michaels's description of the satellite temperature records in the Southern Hemisphere accurate?*

MAHLMAN: As shown above, Michaels's representation of the MSU data set is inappropriate and misleading as an index of global climate trends for comparison to model-calculated trends. Thus, a claim of attribution of specific aspects of Southern Hemisphere behavior, if anything, is even less scientifically rational.

Q: *How would you characterize the model projections of temperature changes in this part of the atmosphere versus the satellite temperature records from the Northern and Southern hemispheres?*

MAHLMAN: As stated above, the twenty-year satellite record is too short to make specific attributions with confidence. Hansen et al. (1995) have performed detailed analyses on the long-term temperature-monitoring value of the MSU data, while Santer et al. (1996) have carefully examined the relationship of observed "trends" versus model predictions that include such complications as the sulfate aerosol effect. The Santer et al. (1996) work reaches a much more positive conclusion about model predictions than is implied in Michaels's testimony. Unfortunately, Michaels does not yet appear to have published his conclusions in the scientific literature. This makes a more careful analysis of his reasoning very difficult.

Q: *Dr. Michaels stated there is an even greater discrepancy between model projections of warming and observed warming in the Northern Hemisphere than in the Southern. He further stated these climate models project very large warming in the polar region, yet the observed warming has been only slight and occurs prior to 1940, which is before greenhouse gas concentrations changed very much. Since then there has been very little observed change, according to Dr. Michaels.*

MAHLMAN: Michaels seems to have missed the point that finding the projections of large polar warming from current data is greatly complicated by the fact that the current warming signal is expected to be rather small, while the northern polar region is dominated by *large* decadal natural variability. This is a bad place to look for a greenhouse warming "smoking gun" or, in Michaels's quest, a lack thereof. Even the Northern Hemisphere as a whole almost certainly has sulfate-aerosol-cooling offset that needs to be quantified carefully before Michaels-type assertions can be properly evaluated.

Q: *Is the temperature record that Dr. Michaels referred to complete enough for reliable analysis?*

MAHLMAN: I don't think the MSU data are good enough or long enough to warrant the kinds of conclusions that Michaels is reaching. Clearly, if one is interested in using climate data to reinforce one's preconceived notions, recent data sets with short duration and significant measurement uncertainties can provide ample ammunition. Such approaches, however, are the very *opposite* of sound, diagnostic science.

Q: *Dr. Michaels stated that Dr. John Mitchell's new climate model suggests a new, lower estimate of climate sensitivity to a doubling of* CO_2*. Do you agree with that assessment? What are the reasons that the IPCC projections of global warming appear to have come down somewhat?*

MAHLMAN: No. I do not agree with Michaels's statement. Michaels's statement appears to be factually incorrect for two important reasons.

First, Dr. John Mitchell's calculations indicate the measured global warming to date is generally consistent with a climate model that calculates a 2.5-degree-C warming response to a doubled CO_2, when the best *guess* value for the current sulfate offset is included.

Second, such a result, in itself, has *nothing* to do with a changed sensitivity to doubled CO_2 in his model, as incorrectly claimed by Michaels.

The sulfate offset of greenhouse-gas-induced warming is the reason for the lowered IPCC warming projections. Michaels and a few others seem to think that a cooling offset somehow lowers the sensitivity of the climate to increased greenhouse gases. I cannot find any logic in such an assertion.

Q: *Dr. Michaels's chart purports to show that all of the temperature change from 1965 to 1994 occurred in one year, a feature that models cannot predict.*

MAHLMAN: This is clearly incorrect. I do not know what method, if any, Michaels used to make this assertion.

Q: *Dr. Michaels testified that the climate models most heavily cited by the IPCC 1992 supplementary report on climate change were known to contain large errors at the time of adoption of the Framework Convention on Climate Change and that such errors were not disclosed, with the result that the model's uncertainties were not considered in the debate surrounding this issue.*

MAHLMAN: I simply do not accept the conspiratorial tone of Michaels's assertions about the IPCC process. First, imperfections in the models are widely and openly acknowledged. . . . Michaels's assertion of a virtual IPCC cover-up is totally implausible, given the very open style of this process.

Contrary to Michaels's assertion, global-scale features in the leading climate models are simulated rather well. Many smaller details, however, exhibit significant disagreements with climate observations. Those problems of detail, however, do not seem likely to be the dominant cause of error in the climate model projections. It still is the large uncertainty in the treatment of cloud feedbacks in the climate system that dominates the current generous error estimates.

Q: *Dr. Michaels testified that the skeptics have been saying for the last seven years that by the year 2100, with a doubling of carbon dioxide, the net warming would be 1 to 1.5 degrees C—a smaller increase than is projected by early climate models. What was the quantitative basis on which the skeptics based such projections of net warming, prior to recent models that incorporate the offsetting influence of aerosols? Did the skeptics conduct a series of model projections? Have the skeptics been able to construct a climate model that, based on rigorous parameterizations, reproduces their projected results?*

MAHLMAN: It is not easy to answer this question directly. Unfortunately, almost none of the skeptics' assertions about low climate sensitivity to increased greenhouse gases have appeared in the open literature. Given below is my best attempt to decipher their assertions:

Skeptics' Assertion 1: "The models are imperfect, so their projections are likely to be overestimated."

Truth: The best estimates that scientists make are designed to equalize the odds that their best guess is either too high or too low.

Skeptics' Assertion 2: "The climate system is resilient and resistant; thus the effects of human beings are likely to be very small."

Truth: The infrared absorption properties of the greenhouse gases being added to the earth's atmosphere are very well known and show that significant perturbations of the climate are likely. The above assertion is not based upon quantitative scientific reasoning.

Skeptics' Assertion 3: "Water vapor will provide a strong negative feedback in the upper troposphere, thus lowering the climate's sensitivity to increased CO_2." This hypothesis, offered in different forms and in informal media by Richard Lindzen, has a physically plausible argument behind it and is partially testable with real atmospheric data. Essentially, the hypothesis is that greenhouse warming of the earth's surface may introduce more deep convection, which could have the effect of drying out the upper troposphere. If this hypothesis were proven to be real, it would have the important effect of reducing the sensitivity of climate to increased CO_2.

Truth: Unfortunately, all available observational tests to date refute this hypothesis. Whatever convection occurs in an area, whether on weather, season, or El Niño time scales, the observed effect is a *moistening* of the upper troposphere. These tests agree, in essence, with the climate model results, thus increasing confidence in their predictions of an amplifying water vapor feedback on the climate sensitivity.

Q: *Dr. Michaels testified that in 1992, in association with the signing of the Rio treaty, Congress was knowingly misled by witnesses who withheld information regarding known errors in key models. Please respond to that statement.*

> MAHLMAN: Michaels's assertion seems to be that the modelers secretly knew that the observed trends were already inconsistent with the model projections. Michaels's perception of the inconsistencies are apparently based upon his conspicuously flawed comparison of a GFDL-idealized CO_2 warming scenario with the record from the MSU satellite data over the past twenty years.

REBUTTALS TO MICHAELS'S CRITIQUE OF BENJAMIN SANTER

In the July 4, 1996 issue of *Nature,* Benjamin Santer, K. E. Taylor, Tom M. L. Wigley, and ten other researchers published an article that concluded: "The observed spatial patterns of temperature change in the free atmosphere from 1963 to 1987 [the period for which various sets of data had been calibrated and integrated] are similar to those predicted by state-of-the-art climate models incorporating various combinations of changes in carbon dioxide, anthropogenic sulfate aerosol and stratospheric ozone. The degree of pattern similarity between models and observations increases through this period. It is likely that this trend is partially due to human activities, although many uncertainties remain, particularly relating to estimates of natural variability."

A *Nature* editorial that accompanied the research concluded: "Despite [several] caveats, the results of Santer et al.—using the available data and state-of-the-art climate models—provide the clearest evidence yet that humans may have affected the global climate."

On July 15, 1996, at an international negotiating conference in Geneva, Patrick Michaels released an article, titled "New Data Cast Doubt on Human Fingerprint," that criticized the *Nature* article. Michaels wrote: "The apparent finding of the human 'fingerprint' on climate caused by the combination of sulfate aerosols and greenhouse gases has been hailed as the most important new scientific information of the year concerning global warming. Particular attention has been paid to the dramatic findings of Santer et al. published in the July 4 issue of *Nature* Magazine which demonstrates an increasing pattern of correlation between two climate models and global observations of weather balloon data from 1963 to 1987.

"It is well known (but not generally acknowledged) that the Lawrence Laboratory model that is used has only ½ of the known greenhouse effect changes, and that when the full effect is put in, the model produces dramatic warmings of several degrees that have simply not been observed. But perhaps more important is the fact that the 'fingerprint' model, and its finding of the greatest warming in the midlatitude Southern Hemisphere—which fits the 'sulfate + greenhouse' models now in vogue—is dramatically changed when all of the upper atmospheric data are included.

"Our picture below shows the years used in Santer et al. in their July 4 'fingerprint' paper. The data are the upper atmospheric record of Angell et al., one of the most highly cited data sets in the refereed climatological literature. Santer et al. used a different record, published by Oort et al., but the two are virtually the same; the correlation between them in the 1963–1987 study period is .92 over the Southern Hemisphere. The difference is that the Angell et al. record is longer, running from 1957 through 1995.

"Note that the years used by Santer et al. to describe the 'fingerprint' coincide with a highly significant warming trend over the region that contributes most to the 'fingerprint.' However, when *all* of the data of Angell are used, *there is no trend whatsoever in the data.* How this glaring problem could have evaded the review process is a legitimate question and must be resolved before the 'fingerprint' studies, such as those in IPCC Chapter 8, serve as the basis for policy."

Michaels's critique of the Santer et al. findings drew a number of rebuttals from leading scientists, among them Michael MacCracken and Tom M. L. Wigley.

A Few Preliminary Comments on the Note by Patrick Michaels entitled "New Data Cast Doubt on Human Fingerprint," by Michael MacCracken

The note by Patrick Michaels entitled "New Data Cast Doubt on Human Fingerprint" suggests that the vertical pattern analysis just published by Santer et al. in *Nature* (July 4, 1996) has selectively used the data (focusing only on a generally warming period from 1963–1987) and ignores a longer data set that, Michaels claims, would give a different result. He thus questions the IPCC conclusion that there has been a discernible human influence on global climate.

I would offer the following conclusions:

1. The Santer et al. study is based on an analysis of the time-evolving vertical-latitudinal pattern of observed temperature changes, meaning that it is focused on how one region of the atmosphere is changing with respect to another. Thus, looking at the time-evolving change in one region (as Michaels does) without looking at how the temperature is changing in other regions (as Santer et al. did) provides no interpretive value with respect to the pattern analysis.

a. The Angell data used by Michaels are an average over the troposphere and do not provide information on the vertical structure. In the Santer et al. analysis, it is this structure that is critical;

b. In that the stratosphere is cooling very strongly, even a slight cooling in the troposphere still means that the stratosphere is cooling relative to the troposphere; and,

c. Not showing what is happening in the Northern Hemisphere does not allow consideration of the point that Santer et al. made that the Southern Hemisphere was warming with respect to the Northern Hemisphere.

2. The Santer et al. study used the Oort data base rather than the Angell data base that Michaels cites. While the Angell data base is useful, it was not suitable for this study for several reasons:

a. The Angell data base does not provide the horizontal and vertical resolution needed to do pattern studies;

b. The early years of the data base (prior to 1963) are very limited in their representation of the Southern Hemisphere and there are known instrument intercalibration problems;

c. The Angell data set is from a fairly limited set of radiosonde stations and so is not as comprehensive spatially as is the Oort data set; and

d. In contrast to the Angell data set, the Oort data set has been carefully reviewed for internal consistency and quality and has been carefully interpolated spatially. This careful analysis has not yet been extended past 1987 (despite requests that it be done) and this is why the Santer et al. analysis stopped when it did. The National Oceanic and Atmospheric Administration/National Climatic Data Center is now preparing a new, carefully checked data set that will likely be used in the next analyses, as is David Parker at the Hadley Centre for Climate Prediction and Research (Bracknell, U.K.).

3. The Santer et al. analyses were not based on a single model, but used the results of model simulations by Taylor and Penner and by Mitchell et al. These results gave confirmatory results:

a. Santer et al. considered a range of possible ratios of the forcing used by Taylor and Penner, so the argument put forth by Pat Michaels is not correct. In their COMB3 sensitivity study, they tried to address the issue of the signal uncertainty introduced by a possible overestimate of the "effective" aerosol forcing. They did this by scaling the aerosol response by 0.5 and leaving the CO_2 response unchanged. They found that the significance levels for the results for the interhemispheric asymmetry component of the signal were sensitive to uncertainties in the ratio of effective GHG [greenhouse gas]/effective aerosol forcing. Crucially, however, their significance results for temperature changes over the full atmospheric profile were not sensitive to these uncertainties! If they had used equivalent CO_2, rather than CO_2 only, their results would have been even more significant—the stratosphere would have shown greater cooling, while the troposphere would have shown greater warming! Pat Michaels's charge to the contrary is incorrect and naive.

b. The Mitchell et al. results were based on a more up-to-date estimate of GHG forcing and, even though a different model, gave the same results. The conclusion is robust.

4. The Michaels note seems to ignore the fact that the Santer et al. analysis was considering multiple human forcings (GHGs, aerosols, and ozone). The Santer et al. analysis did not attempt to include natural forcings (climatic noise), including the rather strong response that can occur following a volcanic eruption (which is exactly counter to a GHG effect—warming the stratosphere and cooling the troposphere), which is almost certainly the reason the Angell data set shows cooling in the 1990s. Thus, the expectation of uniform warming is overly simplified—the expectation is that there should be regions of both warming and cooling and that natural variations can cause changes in trends. In any case, what is important is the pattern analysis—not the trend of a single, spatially integrated variable.

5. The IPCC conclusion that the balance of evidence suggests that human-induced climate change is discernible was based on a number of factors:

a. The global temperature has increased 0.3 to 0.6 degrees C since the mid-nineteenth century. This is confirmed by borehole temperatures, the melting back of mountain glaciers, and rising sea levels (which indicate warming oceans);

b. The present warmth is greater than has prevailed on Earth since at least A.D. 1400, based on paleoclimatic proxy records. The present century is probably the warmest century in a much longer period;

c. The observed global pattern of surface temperature change is generally consistent with GHG and aerosol effects and unlike natural fluctuations;

d. The vertical pattern of change in temperature, which is the subject of this discussion, shows the Southern Hemisphere warming with respect to the Northern Hemisphere and the troposphere warming with respect to the stratosphere. These results are consistent with human influences and unlikely to be due to natural fluctuations; and

e. Model simulations with GHGs and aerosols give a magnitude of change that is generally consistent with observed warming.

Each of these lines of evidence has strengths and shortcomings—that they all occur together was the primary basis for the IPCC conclusion.

Overall, therefore, it appears that Michaels has carried through a seriously flawed analysis indicating poor understanding not only of what Santer et al. did but also how the IPCC came to its conclusion. We also understand that a new and independent analysis of the vertical pattern trend, using data that continue through the end of 1995, will confirm that the observed response is best explained by the combined human factors of changes in GHGs, aerosols, and stratospheric ozone. The Michaels analysis does not in any way diminish the IPCC conclusion regarding the human influence of climate change and the IPCC conclusion remains valid (and thoroughly reviewed, in contrast to the Michaels note.)

Tom M. L. Wigley's Response to Michaels

The detection study to which Michaels refers is based on pattern similarities between observed and model-predicted changes in temperature. Pattern-based studies are independent of the magnitude of area-average changes both in the observations and in the searched-for signal (i.e., the model results.) Michaels's argument, which uses area-average magnitude data, has simply missed the point—trends in these magnitudes are not related to trends in the pattern correlation. It is the latter that demonstrate an increasing similarity between observations and model predictions. Thus, Michaels's arguments are irrelevant and merely expose his ignorance or deliberate misrepresentations of this powerful technique.

Furthermore, IPCC statements regarding detection of an anthropogenic climate change signal do not rest solely, nor even largely, on the recent *Nature* paper to which he refers. They are based on a large body of evidence from diverse sources and from many statistical studies, including other pattern-based studies. This evidence is summarized and evaluated in the extensively peer-reviewed "detection" chapter (Chapter 8) in the IPCC Working Group I Second Assessment Report. Michaels's misguided attempt to shoot down a single swallow will not make the summer go away.

IPCC WORKING GROUP I'S RESPONSE TO MICHAELS

In the Summary for Policy Makers of its Second Assessment Report, published in the spring of 1996, the IPCC's Working Group I also addressed Michaels's findings regarding the evidence for planetary warming.

> The only other recent pattern-oriented work that has attempted to find a CO_2-only signal in the observed surface air temperature data is that by Michaels et al. (1995) This investigation makes use of the time-dependent signal from a transient greenhouse warming experiment performed with the GFDL CGCM (Manabe et al. 1991). The premise underlying this investigation is that if the model-predicted transient signal is not found in the observed temperature record, the model is wrong. The authors fail to find this signal in the observed data, a result that is used to justify a condemnation of climate models in general.
>
> There are a number of serious problems with this analysis. . . . [A] time-dependent greenhouse warming experiment performed with a fully-coupled CGCM does not have a pure signal output. The output consists of signal plus noise, and the early decades of

such simulations are often dominated by the noise. A null result on thc basis of a single transient experiment such as this does not constitute "proof" that the model is erroneous, nor that the searched-for signal does exist.

Furthermore, the Michaels et al. study categorically dismisses the possibility that their failure to find a time-dependent greenhouse-gas signal may be due to the masking effects of anthropogenic sulfate aerosols. This dismissal is made on the following grounds. Michaels et al. argue that if sulfate aerosols have had an impact on climate, then the impact should be very small in regions remote from areas where the forcing due to aerosols is large. This hypothesis is not supported by recent GCM experiments, which suggest that the atmospheric general circulation can, via dynamics, produce large remote surface temperature responses to highly-regionalized forcing by sulfate aerosols (Taylor and Penner 1994; Roeckner et al. 1995; Mitchell et al. 1995).

The Michaels et al. results are difficult to compare with those of other Stage 2 studies that have searched for a CO_2-signal, primarily due to differences in definition of the signal, methodology and in the areas of the globe considered. Nevertheless, their failure to find the sub-global-scale pattern of this signal is consistent with the results of Santer et al. (1993, 1995a). A likely explanation for this result is that some part of the regional-scale features of a CO_2-only signal has been obscured by aerosol effects.

IPCC AUTHORS' RESPONSE TO SINGER

In a letter to *Science* magazine (February 2, 1996) S. Fred Singer charged that the most recent IPCC assessment "presents selected facts and omits important information." His assertions were addressed in a subsequent letter to *Science* (March 15, 1996) by four of the lead authors of the IPCC assessment—Tom Wigley, Benjamin Santer, Dr. J. F. B. Mitchell of the Hadley Centre for Climate Prediction and Research, and Dr. Robert J. Charlson of the University of Washington. The letter was written by Wigley.

The following exchange juxtaposes assertions by Singer, with the specific responses from the IPCC authors.

SINGER: The summary (correctly) reports that the climate has warmed by 0.3 to 0.6 degrees C in the last 100 years, but does not mention that there has been little warming . . . in the last 50 years, during which time some 80 percent of greenhouse gases were added to the atmosphere.

RESPONSE: [Singer's assertion] is not supported by the data; the warming from 1946 to 1995 is 0.3 degrees C. As shown in chapter 8 of the full [1995 IPCC] report, there is no inconsistency between the observed temperature record and model simulations.

SINGER: With climate models lacking in validation, why . . . should we trust any of the forecasts about future warming, sea level rise, and other claimed impacts—or use them as the basis for costly policies?

RESPONSE: Singer writes that climate models lack validation. Chapter 5 of the [1995 IPCC] report deals with the validation issue. Current general circulation models (GCMs) have well-known and well-documented weaknesses, but they still perform remarkably well in simulating important features of our current climate conditions.

SINGER: The IPCC summary does not mention explicitly that—thanks to the inclusion of previously neglected aerosols in global circulation models (GCMs)—its 1995 temperature forecasts are one third less than the range of values endorsed just three years ago. Yet statesmen signing a Global Climate Treaty in Rio, including George Bush, were assured that the IPCC forecasts represented a "scientific consensus" and were "of the highest quality."

RESPONSE: In noting that global mean temperature projections made in the latest assessment are lower than those given in 1992, Singer says this fact and the reasons for it are not mentioned in the [IPCC] Summary for Policy Makers.

This is incorrect. There are a number of reasons for the differences, which are given in the [Summary] and in the relevant chapters of the full report.

SINGER: The cooling effects of aerosols have been well recognized for some 30 years and have been invoked by climate scientists . . . to explain the climate cooling observed between 1940 and 1975. Yet aerosols were incorporated into GCMs only recently—and only imperfectly. Man-made aerosols encompass a wide variety of particulates—sulfates from the emissions of SO_2 in fossil fuel combustion to smoke and soot from forest clearing and other biomass burning. Because these have quite different optical properties, their climate effects will also be quite different. . . .

To the extent that pollution control by major emitting nations is reducing the creation of sulfate aerosols, one would expect the current average warming to be greater than 0.3 degree C per decade, and one would expect to see enhanced regional differences, making the disagreement with observations even greater.

RESPONSE: Singer notes that potential aerosol influences on climate were hypothesized some 30 years ago. Current literature recognizes these early empirical works. What has changed since 1990 has been our ability to quantify the influences of specific chemically defined types of aerosols and include them as independently estimated forcings. The IPCC report considers both direct and indirect effects of anthropogenic sulfate and carbonaceous aerosols and provides estimates of uncertainties. Further, it documents current increases in world sulfur emissions (mainly from Asia), contrary to Singer's implication that pollution controls by major emitting nations might have caused global decreases.

SINGER: In view of the above, it is difficult to give credence to the statement [of the IPCC] that "over recent decades the observed spatial pattern of temperature change increasingly resembles the expected greenhouse-aerosol pattern." The research has not yet, to my knowledge, appeared in the peer-reviewed literature, violating a major rule of the IPCC. More important, there has not been time for an independent scrutiny to see, for example, whether the resemblance really "increases," irrespective of the GCM and aerosol scenarios that are used.

RESPONSE: [Singer] is wrong on both counts. The criterion for inclusion of material in the IPCC reports is not that the material should be in the peer-reviewed literature, but that it should be accessible to reviewers of IPCC drafts. . . . The reason for this is partly to ensure that the report, when published, would be up-to-date and truly reflect the state of the art. In any event, more than 95 percent of the work cited in chapter 8 *is* in the peer-reviewed literature.

The specific work Singer refers to, on the increasing correlation between the expected greenhouse-aerosol pattern and observed temperature changes, *is* in the peer-reviewed literature. Furthermore, the former work has been available since January 1995. Other work on this topic that is cited in chapter 8 is also readily available. . . . Our plea to Singer and others who comment on IPCC reports is that they show the same concern for accuracy and balance as do those scientists who worked so hard to prepare the IPCC reports and who assisted in the review and approval process.

SINGER-SANTER EXCHANGE IN *THE WALL STREET JOURNAL*

The following letter by S. Fred Singer was published in *The Wall Street Journal* on July 25, 1996. Singer wrote the letter in response to the Santer findings published in *Nature* on July 4, 1996.

Dr. Benjamin Santer was the convening lead author of Chapter 8 of the 1995 UN scientific report prepared by the Intergovernmental Panel on Climate Change (IPCC).

This is the crucial chapter supporting the rather feeble IPCC conclusion that "the balance of evidence suggests that there is a discernible human influence on global climate." In the absence of any evidence for a current warming trend, this artful phrase is being used to frighten politicians into believing that a climate catastrophe is about to happen. The State Department has evidently adopted this view and—at the just concluded UN conference in Geneva—proposed that mandatory restrictions on energy use replace the voluntary plan now in effect.

Chapter 8 is based mainly on two research papers, both co-authored by Santer, one published in *Climate Dynamics* in December 1995, and the other in the July 4, 1996, issue of *Nature.* The draft of Chapter 8, under Santer's direction, was mailed for comment in May 1995, well before either paper was published and available for critical evaluation. Eight of Santer's co-authors on these papers are also listed as contributors to Chapter 8. It is really too much to ask of any scientist that he effectively criticize his own research. Instead, the IPCC should be faulted for permitting the lead author of a crucial chapter to use his own unpublished work as a basis and for not including as a lead author my University of Virginia colleague Professor Patrick J. Michaels, who, at the time, had published the only refereed paper on the subject.

We have now had a chance to examine Santer's research papers and note the following:

> 1. The conclusion about the "human influence" depends on the correlation (between the geographic patterns of observed and calculated temperature changes) showing an increasing trend in time. Indeed, Figure 8.10(b) in Chapter 8 exhibits a straight-line, positive trend over the last 50 years. This figure is taken directly from Santer's *Climate Dynamics* paper; but there it shows also negative trends. The trend over the last 20 years is negative; the trend over the last 25 years is zero. In other words, the trend depends entirely on how one selects the time interval. What is most disturbing, however, is the fact that someone has edited out the other (four) trends displayed in Santer's original figure, leaving the reader of the IPCC report with the impression that only a positive trend exists.

I have twice questioned Santer about this clear instance of "scientific cleansing," but so far he has not replied.

2. Another pattern correlation used to support the IPCC conclusion depends on the existence of a large increasing temperature trend in the Southern Hemisphere mid-troposphere, shown clearly in Santer's *Nature* paper and in Chapter 8, Figure 8.7(c). But this "positive trend" exists only for the data set covering the years 1963–1987, which was used by Santer et al., in their analysis. The trend disappears completely, and becomes a zero trend, if one uses the full data set available, from 1957 to 1995. Pat Michaels has sent a letter to *Nature* covering this selective use of data.

I am not alleging here that Santer consciously selected the data to match the desired conclusion. I only claim that his conclusion about an increasing correlation trend does not hold up under scientific scrutiny, and neither does the IPCC conclusion about a "discernible human influence," based on Santer's work.

(Signed) Dr. S. Fred Singer

On July 26, 1996, Benjamin Santer wrote this unpublished reply to Singer:

Dr. S. Fred Singer's letter of July 25 to *The Wall Street Journal* contains serious factual inaccuracies.

1. Singer states that there is "an absence of any evidence for a current warming trend." This is untrue. There is clear evidence that the Earth has warmed by 0.3–0.6 degrees Celsius over the past century. This evidence is thoroughly documented in Chapter 3 of the 1995 IPCC Working Group I Report on the "Science of Climate Change." Anyone who could read this report and reach the conclusion that there is "an absence of any evidence for a current warming trend" is seriously out of touch with reality.

2. Singer claims that "Chapter 8 is mainly based on two research papers." Again, this erroneous statement serves

to document Singer's inability to argue on the basis of the facts. Chapter 8 references more than 130 scientific papers—not just two. Its bottom-line conclusion that "the balance of evidence suggests a discernible human influence on global climate" is not solely based on the two Santer et al. papers that Singer alludes to. This conclusion derives from many other published studies on the comparison of modeled and observed patterns of temperature change—for example, papers by Karoly et al. (1994), Mitchell et al. (1995), Hegerl et al. (1995), Karl et al. (1995), Hasselmann et al. (1995), Hansen et al. (1995), and Ramaswamy et al. (1996). It is supported by many studies of global-mean-temperature changes, by our physical understanding of the climate system, by our knowledge of human-induced changes in the chemical composition of the atmosphere, by information from paleoclimatic studies, and by a wide range of supporting information (sea level rise, retreat of glaciers, etc.). To allege, as Singer does, that "Chapter 8 is mainly based on two research papers" is just plain wrong.

3. Singer further alleges, "The draft of Chapter 8, under Santer's direction, was mailed for comment in May 1995, well before either paper was published and available for critical evaluation." Again, this is demonstrably untrue. Drafts of both the Santer et al. papers that Singer refers to were available to reviewers upon request during the country review stage. No such request was made by Dr. Singer. One of the Santer et al. papers was published as PCMDI report #21 in January 1995—five months before the start of the IPCC country review process. (PCMDI reports are routinely distributed to over 300 scientists and scientific organizations worldwide). Finally, as noted by Tom Wigley, John Mitchell, Bob Charlson, and myself in a letter to *Science* (March 15, 1996):

"The criterion for inclusion of material in the IPCC reports is not that the material should be in the peer-reviewed literature, but that it should be accessible to reviewers of

IPCC drafts. Thus, published reports, book chapters, and manuscripts submitted for publication or in the press, were acceptable material. The reason for this is partly to ensure that the report, when published, would be up-to-date and truly reflect the state of the art."

4. Singer bemoans the failure of the IPCC "for permitting the lead author of a crucial chapter to use his own unpublished work as a basis and for not including as a lead author my University of Virginia colleague Professor Patrick J. Michaels, who, at the time, had published the only refereed paper on the subject." Here, too, Singer seems to be viewing reality through some strange distortion filter. The implication of his statement is that none of my work has been published in the peer-reviewed literature, and that Pat Michaels published the first paper on pattern-based detection. I doubt whether even Pat Michaels would make such outrageous claims.

First, had Singer read Chapter 8 thoroughly, he would have noted that Chapter 8 cites two pattern-based detection studies (by myself and other scientists) that appeared in a peer-reviewed scientific publication, *Climate Dynamics,* in 1993 and 1994. By my reckoning, 1993 and 1994 predate the start of the IPCC country review process in May 1995.

Second, Pat Michaels was not the first to publish refereed papers on the subject of pattern-based detection. The first practical pattern-based study was by Tim Barnett and Michael Schlesinger in 1987. This was published in the peer-reviewed *Journal of Geophysical Research* (an earlier theoretical paper on this subject was published by Klaus Hasselmann in 1979). Pat Michaels's pattern-based study was published seven years *after* the Barnett and Schlesinger paper. It appeared in *Technology: Journal of the Franklin Institute.* The review process could not have been that arduous: the paper was received on September 7, 1994, and accepted for publication on October 10, 1994.

Third, Pat Michaels was invited to contribute to Chapter 8. He declined to do so. One of the lead authors of Chapter 8, Tom Wigley, wrote to Pat Michaels on November 21, 1994, and on February 21, 1995, soliciting comments on the portrayal of Michaels's Franklin Institute paper in a December 8, 1994 version of Chapter 8. Prof. Michaels did not respond to these requests.

5. Singer claims that Chapter 8 has been "scientifically cleansed." The specific act of cleansing that he refers to pertains to Figure 8.10b in Chapter 8. This figure shows the correlation between a model-predicted pattern of temperature change (in response to combined changes in CO_2 and sulfate aerosols) and observed patterns of temperature change over 1910 to 1993. The figure also includes a least-squares linear trend fitted to the correlation time series over the 50-year period 1944–1993. Singer decries the fact that other, shorter timescale trends are not shown on this figure, although they are shown in our 1995 *Climate Dynamics* paper (Figure 10). The shorter timescale trends are nonsignificant. Singer implies that the IPCC report fails to show these shorter timescale trends simply because they are nonsignificant, and hence that I have suppressed scientific information that would tend to contradict the conclusion that there is a discernible human influence on climate.

Singer's basic problem here is that he does not understand the concepts of signal and noise. On timescales of 10 years or less, ambient noise levels are large (arising from ENSO-type variability, the short-term transient effects of volcanic eruptions, etc.), and signals of human effects tend to be small, since the changes in human-induced radiative forcing tend to be small. Signal-to-noise (S/N) ratios for such short timescale trends are therefore low, and it is difficult to detect human influences. A case in point is the short (17-year) MSU time series. On longer timescales, noise levels tend to decrease, changes in human influences are larger, and hence any anthropogenic signal (and S/N ratios) will be larger. This means that if one wants to search for an anthropogenic signal in observed data, there is little sense in making short timescale trends the focus of such a search! Our 1995 *Climate Dynamics* paper and 1996 *Nature* paper consider the issue of trend significance on a range of timescales. The same is true of the 1996 Hegerl et al. paper, soon to appear in the *Journal of Climate.*

Our papers and Gabi Hegerl's paper provide full information on trend significance as a function of timescale and season. Figures 8.10a and 8.10b were restricted to 50-year timescale trends as a pedagogical example of the differences in model-versus-observed pattern similarity between "CO_2-only" and "CO_2+aerosol" signals. There was and is no sinister plot to suppress uncertainties. In fact, Singer fails to note that Figure 8.3 in the now-published Chapter 8 *specifically illustrates* trend significance as a function of timescale (from 10 to 100 years), and shows that the shortest timescale trends (10-year trends) are nonsignificant! Furthermore, the issue of nonsignificance of short timescale trends is comprehensively covered in Chapter 8:

"The key point to note here is that the MSU record is short ($<$20 years) for the purposes of detecting a slowly evolving anthropogenic signal. This short record limits comparisons of satellite-based and model-predicted data to decadal time-scale temperature trends. These trends are strongly affected by the background noise of interannual to decadal time-scale natural variability (see, for example, the lack of significance for the most recent 10-year trends in near-surface temperature in Figure 8.3). It is therefore difficult to make a meaningful interpretation of any differences in trend on these short time-scales" (page 438).

(Signed) Dr. Benjamin D. Santer

MACCRACKEN'S REVIEW OF BALLING'S *THE HEATED DEBATE*

In 1992 Robert Balling, Jr., director of the Office of Climatology at Arizona State University and a prominent greenhouse skeptic, published a book titled *The Heated Debate,* which significantly downplays climate concerns.

The following observations are excerpted from Michael MacCracken's review of Balling's book in the September 1993 *Bulletin of the American Meteorological Society:*

Balling's book is . . . frustrating. Despite its title, the book is clearly not a documentary of the debate that is taking place. The popular vision [of catastrophic climate change] he describes is neither well referenced nor responsibly filtered, with Balling picking and choosing from the most dire consequences put forth by activist environmental groups (and some scientists in their public statements) to make up the apocalyptic future. There is no effort made to provide the rationale, explanations, justifications, or contexts that these groups offer for basing their conclusions on worst-case scenarios (e.g., taking a precautionary approach that assumes the knowns plus uncertainties will occur in that the effects are long term and irreversible, there is only one "spaceship Earth" on which life is known to be able to survive.) Thus Balling sets up a "straw man" catastrophist vision in which it is rather easy to punch holes. It would have been much more of a challenge had he taken on the more authoritative report of the Intergovernmental Panel on Climate Change (IPCC). . . .

What is frustrating, however, is that Balling seems to have forgotten what he had just described in the preceding chapter; namely, that the climate is indeed about as sensitive as the models are indicating . . . and so, with sulfates having a short atmospheric lifetime and greenhouse gases a long lifetime, the greenhouse effect will prevail over time, and at an accelerating rate. In . . . the final epilogue, Balling eschews use of the "crystal ball," relying instead on conclusions from the past. He criticizes the global warming catastrophists of having a religious attachment to the issue and of not accepting lessons from the climate past; yet Balling seems to be as tightly attached philosophically to future changes being limited to those in the past, even as population growth proceeds exponentially.

NOTES

INTRODUCTION

pages 1–3: Descriptions of Larsen ice shelf fracture from reports by Reuters.

page 3: "Comparable temperature changes . . . : George C. Marshall Institute, *The Global Warming Experiment,* 1995.

page 4: "I am tempted to ask . . . : Dana Rohrabacher, at House Science Committee, Subcommittee on Energy and Environment, *Hearing on Scientific Integrity and the Public Trust: Case Study 2—Climate Models and Projections of Potential Impacts of Global Climate Change,* 104th Cong., 1st sess., November 16, 1995, Report no. 31.

page 5: 2,500 leading climate scientists . . . : IPCC Working Group II, *Summary for Policymakers: Scientific Technical Analysis of Impacts, Adaptations and Mitigation of Climate Change,* November 1995.

page 8: the ten hottest years in recorded history . . . : "'95 the Hottest Year on Record As the Global Trend Keeps Up," *New York Times,* January 4, 1996.

page 8: "If the last 150 years had been marked . . . : James J. McCarthy, interview by author, Cambridge, Mass., June 20, 1995.

page 10: Insurance industry data from Munich Re/Münchener Rückversicherungs-Gesellschaft, in Christopher Flavin and Odil Tunali, *Climate of Hope: New Strategies for Stabilizing the World's Atmosphere* (Worldwatch Institute, June 1996).

page 10: In 1995 Andrew Dlugolecki . . . : Jeremy Leggett, ed., *Climate Change and the Financial Sector: The Emerging Threat—The Solar Solution* (Munich: Gerling Akademie Verlag, 1996), p. 64.

CHAPTER ONE: OF TERMITES AND COMPUTER MODELS

page 15: "Termites are everywhere . . . : "When Winters Go Frost-Free, It's Bug Easy," *Boston Globe,* May 29, 1995.

page 15: the fourth year of the worst drought . . . : "Withering Drought Hits Parts of Spain," Reuters, June 14, 1995.*

page 15: Russian thermometers soared . . . : "For Moscow, Tables Turned: Record Heat," *Boston Globe,* June 2, 1996.

page 16: unseasonal torrential rainstorms . . . : "China Floods Kill Nearly 1,200," Associated Press, July 2, 1995; remarks of Lin Erda from New China News Agency, June 11, 1995, BBC Summary of World Broadcasts, June 13, 1995.

page 17: The IPCC Working Group I (on science) has published four reports—in 1990, 1992, 1994, and 1995. The IPCC Working Group II (on impacts, adaptations, and mitigation of climate change) has published one report in 1995. The IPCC Working Group III (on the social and economic dimensions of climate change) has published one report in 1995. In addition, the IPCC published a synthesis report in 1996:

The 1990 Assessment: *Climate Change: The IPCC Scientific Assessment,* J. T. Houghton, G. J. Jenkins, and J. J. Ephraums (eds.). 1990. Cambridge, U.K.: Cambridge University Press.

The 1992 Assessment: *Climate Change 1992: The Supplementary Report to the IPCC Scientific Assessment,* J. T. Houghton, B. A. Callander, and S. K. Varney (eds.). 1992. Cambridge, U.K.: Cambridge University Press.

The 1994 Report: *Climate Change 1994: Radiative Forcing of Climate Change and An Evaluation of the IPCC IS92 Emission Scenarios,* J. T. Houghton, L. G. Meiro Filho, J. Bruce, Hoesung Lee, B. A. Callander, E. Haites, N. Harris, and K. Maskell (eds.). 1994. Cambridge, U.K.: Cambridge University Press.

The 1995 Report: *Climate Change 1995—The Science of Climate Change: Contribution of Working Group I to the Second Assessment Report of the Intergovernmental Panel on Climate Change,* J. T. Houghton, L. G. Meira Filho, B. A. Callander, N. Harris, A. Kattengberg, and K. Maskell (eds). 1996. Cambridge, U.K.: Cambridge University Press.

The 1995 Report: *Climate Change 1995: Impacts, Adaptations and Mitigation of Climate Change: Scientific-Technical Analysis: Contribution of Working Group II to the Second Assessment Report of the Intergovernmental*

* Unless otherwise noted, all stories from Reuters News Service and UPI were provided by the ClariNews on-line news wire service.

Panel on Climate Change, Robert T. Watson, Marufu C. Zinyowera, and Richard H. Moss (eds). 1996. Cambridge, U.K.: Cambridge University Press.

The 1995 Report: *Climate Change 1995: Economic and Social Dimensions of Climate Change: Contribution of Working Group III to the Second Assessment Report of the Intergovernmental Panel on Climate Change,* James P. Bruce, Hoesung Lee, and Erik F. Haites (eds). 1996. Cambridge, U.K.: Cambridge University Press.

The 1995 Synthesis: *Climate Change 1995: IPCC Second Assessment Synthesis of Scientific-Technical Information Relevant To Interpreting Article 2 of the UN Framework Convention On Climate Change 1995,* 1996. Cambridge, U.K.: Cambridge University Press.

page 18: "we are in uncharted waters . . . : James J. McCarthy, interview by author, Cambridge, Mass., June 20, 1995.

page 18: more than 70 people died . . . : "Death Toll in Bangladesh Floods Reaches 71," Reuters, June 23, 1995.

page 19: Ghana experienced its heaviest . . . : "This Is Global Warming?" *Newsweek,* January 22, 1996.

page 19: northeastern Brazil suffered its worst drought . . . : "Mudslides Kill at Least 26 Brazilians," UPI, April 23, 1996.

page 20: the planet's surface air temperature has increased . . . : J. Hansen, R. Ruedy, M. Sato, and R. Reynolds, "Global Surface Air Temperature in 1995: Return to Pre-Pinatubo Level," forthcoming in *Geophysical Research Letters.*

page 21: enormous fires burning in Canadian . . . : study by Canadian Forest Service, Northern Forestry Centre, reported by Greenpeace, Canada Press Release, June 16, 1995.

page 21: western portions of Australia . . . : "This Is Global Warming?" *Newsweek,* January 22, 1996.

page 22: "A pattern of climatic response . . . : IPCC Working Group I, *Summary for Policymakers,* November 1995.

pages 22–23: "There is no debate among . . . : James J. McCarthy, interview by author, Cambridge, Mass., June 20, 1995.

page 23: Information on Typhoon Angela, record cold in Mexico, and record snowfalls in Sapporo, from "This Is Global Warming?" *Newsweek,* January 22, 1996.

page 24: a team of scientists at the NOAA's . . . : Thomas R. Karl, Richard W. Knight, David R. Easterling, and Robert G. Quayle, "Trends in U.S. Climate During the Twentieth Century," *Consequences* (Spring 1995). See also "Trends in High-frequency Climate Variability in the Twentieth Century," *Nature,* vol. 377 (September 21, 1995).

page 24: grass and brush fires blackened . . . : "Texas Heat Contributes to Fires

That Destroy 65 Homes," *New York Times,* February 23, 1996; "Texas Wildfires Burn; No Rain Forecast," UPI, March 20, 1996.

page 24: a deadly blizzard . . . : "Blizzard Spreading Famine in Western China Provinces," Reuters, March 5, 1996.

page 25: more than 20 percent of Laos's rice paddies . . . : "Laotians Hit by Rice Shortage after 1995 Floods," Reuters, March 23, 1996.

page 25: As late as May 13 . . . : *Boston Globe,* May 14, 1996; *Boston Globe,* May 23, 1996.

page 26: "what if the science is wrong? . . . : James Watkins, interview by author, offices of *The Boston Globe,* 1991.

page 27: At a concentration thirty times greater . . . : Michael Oppenheimer and Robert H. Boyle, *Dead Heat: The Race Against the Greenhouse Effect* (New York: Basic Books, 1990).

page 27: A more dramatic feedback . . . : "In Alaska's Northern Tundra, Scientists Find Cause for Concern," *Boston Globe,* March 15, 1993.

page 28: "The inhabitants of planet Earth . . . : Wallace Broecker quoted in the Atmosphere Alliance's *Journal of Grassroots Action to Protect the Atmosphere,* No Sweat News, Olympia, Washington (Summer 1996).

page 29: A third important contribution . . . : David J. Thomson, "The Seasons, Global Temperature and Precession," *Science,* vol. 268 (April 7, 1995).

page 29: spring is now arriving a week earlier . . . : C. D. Keeling, J. F. S. Chin, and T. P. Whorf, "Increased Activity of Northern Vegetation Inferred from Atmospheric CO_2 Measurements," *Nature,* vol. 382 (July 11, 1996).

page 30: researchers examining them found . . . : Scott J. Lehman and Lloyd D. Keigwin, "Sudden Changes in North Atlantic Circulation During the Last Deglaciation," *Nature,* vol. 356 (April 30, 1992).

page 30: "You don't want to push your luck . . . : Scott J. Lehman, "Ice Cores Raise the Question: Is Our Luck Running Out?" *Boston Globe,* July 19, 1993.

page 31: "shutdown or comparable drastic change . . . : Wallace S. Broecker, "Chaotic Climate," *Scientific American,* November 1995.

page 31: "When researchers finally conducted actual . . . : Author's notes from seminar sponsored by the Environmental Health Center of the National Safety Council at Tufts University, March 18, 1995.

CHAPTER TWO: THE BATTLE FOR CONTROL OF REALITY

page 34: "Persuasion by its definition . . . : Quoted in John Stauber and Sheldon Rampton, *Toxic Sludge is Good For You!,* (Monroe, Me.: Common Courage Press, 1995).

page 34: The Information Council on the Environment . . . : ICE documents in author's possession, Mission Statement, Strategy Statement, and Test Market Proposal, February 1991.

page 35: I co-authored an article . . . : "Should We Fear a Global Plague?" *Washington Post,* March 19, 1995, Outlook section.

page 36: "there has been a close to universal impulse . . . : Western Fuels annual reports, 1992, 1993, 1994.

page 36: Western Fuels spent $250,000 to produce a video . . . : Idso testimony, at Senate Committee on Commerce, Science and Transportation, *Hearing on Global Change Research: Global Warming and the Biosphere,* 102nd Cong., 2nd sess., April 9, 1992.

page 36: John Sununu's favorite movie . . . : *The Greening of Planet Earth* © Western Fuels Assn., 1991.

page 37: a panel of the World Health and World Meteorological organizations . . . : "U.N. Agencies Say Warming Poses Threat to Public Health," *New York Times,* July 8, 1996; Paul R. Epstein, interview by author, Cambridge, Mass., March 1, 1995.

page 38: enhanced CO_2 would be devastating . . . : Cynthia Rosenzweig and Daniel Hillel, "Potential Impacts of Climate Change on Agriculture and Food Supply," *Consequences,* vol. 1, no. 2 (Summer 1995).

page 38: a report by the private weather-forecasting firm . . . : Accu-Weather report "Changing Weather? Facts and Fallacies About Climate Change and Weather Extremes" (1995).

page 39: Called to the stand . . . : Lindzen, Michaels, and Balling statements at the Minnesota Public Utilities Commission, *Before the Office of Administrative Hearings of the State of Minnesota: In the Matter of the Quantification of Environmental Costs Pursuant to Laws of Minnesota, 1993,* chap. 356, sec. 3, vols. 8, 9, 10, May 22, 23, and 24, 1995. Transcript of proceedings in author's possession.

page 40: "It came out pretty clearly . . . : Quoted in Willett Kempton, James S. Boster, and Jennifer A. Hartley, *Environmental Values in American Culture* (Cambridge, Mass.:MIT Press, 1995).

pages 40–41: Pat Michaels revealed under oath . . . : Rebuttal testimony of Patrick J. Michaels, *Before the Minnesota Public Utilities Commission: In the Matter of the Quantification of Environmental Costs Pursuant to Laws of Minnesota, 1993,* chap. 356, sec. 3, March 15, 1995.

page 42: "The fact is that the artifice . . . : "Forging a Scientific Consensus," *World Climate Review,* vol. 3, no. 1 (Fall 1994), pp. 19, 20.

pages 43–44: Idso's testimony in St. Paul . . . : *Before the Office of Administrative Hearings of the State of Minnesota: In the Matter of the Quantification of Environmental Costs Pursuant to Laws of Minnesota, 1993,* chap. 356, sec. 3, vol. 10, May 24, 1995.

pages 44–45: Balling has also received . . . : Rebuttal testimony of Robert C. Balling, *Before the Minnesota Public Utilities Commission: In the Matter of the Quantification of Environmental Costs Pursuant to Laws of Minnesota, 1993,* chap. 356, sec. 3, March 15, 1995; copy in author's possession.

page 45: Balling disclosed his industry funding . . . : Testimony of Robert C. Balling, *Before the Office of Administrative Hearings of the State of Minnesota: In the Matter of the Quantification of Environmental Costs Pursuant to Laws of Minnesota, 1993,* chap. 356, sec. 3; vol. 10, May 24, 1995.

page 45: "by a few group leaders . . . : Robert Balling, "Keep Cool About Global Warming," *Wall Street Journal,* October 16, 1995.

page 46: Singer did not deny having . . . : "Is Environmental Science for Sale?" ABC News *Nightline,* no. 3329, February 24, 1994.

page 47: Singer's proposed oil-company-sponsored . . . : *A Public Education Program on Global Warming,* proposed by S. Fred Singer; copy in author's possession.

page 47: "the Swedish Academy of Sciences has chosen . . . : S. Fred Singer, "Ozone Politics with a Nobel Imprimatur," *Washington Times,* November 1, 1995.

page 48: "Early this year . . . : S. Fred Singer, "Rays of a Setting Global Warming Sun," *Washington Times,* March 21, 1996.

pages 48–49: According to the World Meteorological Association . . . : "1995 Pegged as Hottest Year on Record in UN Report," *New York Times,* May 1, 1996.

page 49: But Lindzen's theory has been contradicted . . . : "Greenhouse Science Survives Skeptics," *Science,* vol. 256 (May 1992).

page 49: according to satellite readings . . . : B. J. Soden and R. Fu, "A Satellite Analysis of Deep Convection, Upper Tropospheric Humidity, and the Greenhouse Effect," *Journal of Climate,* vol. 8, (1995), pp. 2333–2351.

pages 49–50: Testifying before a Senate committee . . . : Richard S. Lindzen, at Senate Committee on Commerce, Science and Transportation, *The Role of Clouds in Climate Change,* October 7, 1991.

page 51: a team of researchers with the NOAA . . . : S. J. Oltmans and D. J. Hofmann, "Increase in Lower-stratospheric Water Vapour at a Mid-latitude Northern Hemisphere Site from 1981 to 1984," *Nature,* vol. 374 (March 9, 1995).

page 52: Lindzen invited me to his home . . . : Richard S. Lindzen, interview by author, Newton, Mass., August 4, 1995.

page 52: In a striking piece of testimony . . . : Richard S. Lindzen testimony, *Before the Office of Administrative Hearings of the State of Minnesota: In the*

Matter of the Quantification of Environmental Costs Pursuant to Laws of Minnesota, 1993, chap. 356, sec. 3, vol. 9, May 23, 1995.

page 53: "exaggerating risk . . . : Frederick Seitz, letter to U.S. undersecretary of state, Timothy Wirth, September 26, 1994; copy in author's possession.

page 53: "It seems to us . . . : Bert Bolin, letter to Frederick Seitz, November 30, 1994; copy in author's possession.

page 53: from the "scientific backwater . . . : Richard S. Lindzen, interview by author, Newton, Mass., August 4, 1995.

page 54: In his 1992 address to OPEC . . . : Richard S. Lindzen, "Global Warming: The Origin and Nature of Alleged Scientific Consensus," speech delivered to a meeting of OPEC, Vienna, Austria, April 1992; copy in author's possession.

page 55: "The public press," he said . . . : Bert Bolin, *Report to the Eleventh Session of the Intergovernmental Negotiating Committee for a Framework Convention on Climate Change,* New York, February 6, 1995 (emphasis Bolin's).

page 55: "have not been subject to the careful . . . : Bert Bolin, *Report to the Second Session of the Conference of the Parties to the UN Framework Convention on Climate Change,* Geneva, Switzerland, July 8, 1996.

page 55: In response to a series of Freedom of Information Act requests . . . : Funding figures provided by the Internal Revenue Service.

page 56: "If the critical questions about climate . . . : William Ruckelshaus, interview by author, May 9, 1996.

page 58: "the proper role of a scientist . . . : Stephen H. Schneider, "Is the 'Scientist-Advocate' an Oxymoron?" paper presented at the American Association for the Advancement of Science, February 12, 1993.

page 58: "Because I speak with credentials . . . : Jerry Mahlman testimony, House Science Committee, Subcommittee on Energy and Environment, *Hearing on Scientific Integrity and the Public Trust: Case Study 2—Climate Models and Projections of Potential Impacts of Global Climate Change,* 104th Cong., 1st sess., November 16, 1995, Report no. 31.

page 59: "Scientists, by nature, are very . . . : James J. McCarthy, interview by author, Cambridge, Mass., June 20, 1995.

page 60: "The IPCC undermines its scientific . . . : European Science and Environment Forum, "Scientists Attack 'Official Consensus' on Global Warming," press release, March 4, 1996.

page 60: "people were reduced to eating leaves . . . : "North Korea Says Needs More Aid for Flood Relief," Reuters, April 1, 1996.

page 60: the Blue Hills meteorological station . . . : "April 96: One for the Books," *Boston Globe,* April 11, 1996.

page 61: a fresh foot of snow fell . . . : "When Winter's Grip Wouldn't Let Go," *New York Times,* May 19, 1996.

CHAPTER THREE: A CONGRESSIONAL BOOK BURNING

pages 64–67: All congressional testimony and comments from House Subcommittee on Energy and Environment of the Committee on Science, *Hearing on Scientific Integrity and the Public Trust: The Science Behind Federal Policies and Mandates: Case Study 1—Stratospheric Ozone: Myths and Realities,* 104th Cong., 1st sess., September 20, 1995, Report no. 31.

page 67: Whelan, who praises the nutritional value . . . : Howard Kurtz, "Dr. Whelan's Media Operation," *Columbia Journalism Review* (March–April 1990); funders of The American Council on Science and Health are listed in *Chronicle of Philanthropy,* June 12, 1992.

page 67: she attacked a crusade against . . . : Quoted in John Stauber and Sheldon Rampton, *Toxic Sludge is Good For You!* (Monroe, Me.: Common Courage Press, 1995).

pages 67–74: All congressional testimony and comments from House Subcommittee on Energy and Environment of the Committee on Science, *Hearing on Scientific Integrity and the Public Trust: Case Study 2—Climate Models and Projections of Potential Impacts of Global Climate Change,* 104th Cong., 1st sess., November 16, 1995, Report no. 31.

pages 74–76: All congressional testimony and comments from House Committee on Science, *Hearing on Global Change Research Programs: Data Collection and Scientific Priorities,* 104th Cong., 2nd sess., March 6, 1996.

pages 76–78: the Science Committee finally determined what programs . . . : Omnibus Civilian Science Authorization Act of 1996, *Report of the Committee on Science, House of Representatives on H. R. 3322,* May 1, 1996.

pages 78–79: two of the leading IPCC scientists . . . : Author's notes from presentation by Benjamin Santer and Tom M. L. Wigley at a seminar hosted by the U.S. Global Change Research Program, Rayburn House Office Building, Washington, D.C., May 21, 1995.

page 78: Santer, Wigley, and eleven other researchers . . . : B. D. Santer et al., "A Search for Human Influences on the Thermal Structure of the Atmosphere," *Nature,* vol. 382 (July 4, 1996).

page 79: Shortly thereafter, the coal and oil lobbies . . . : Dennis Wamstead, "Doctoring the Documents?" *Energy Daily,* May 22, 1996.

page 80: "I have never witnessed a more disturbing corruption . . . : Frederick Seitz, "A Major Deception on Global Warming," *Wall Street Journal,* June 12, 1996.

page 80: The story—with all its damning . . . : "U.N. Climate Report Was Improperly Altered, Underplaying Uncertainties, Critics Say," *New York Times,* June 17, 1996.

page 80: raising "very serious questions . . . : Patrick Michaels, "Bait and Switch? IPCC Pares Down the Consensus," *World Climate Report,* June 10, 1996.

page 80: In an interview, Santer expressed . . . : Benjamin Santer, interview by author, May 24, 1996.

page 81: In their reply to *Energy Daily* . . . : Editorial, *Energy Daily,* June 3, 1996; copy in author's possession.

page 81: Another letter to the *Journal* . . . : Editorial, *The Wall Street Journal,* June 25, 1996.

page 81: He publicly called for . . . : "Attacks on IPCC Report Heat Controversy Over Global Warming," *Physics Today,* August 1996.

page 81: Rohrabacher wrote to Secretary of Energy . . . : Rep. Dana Rohrabacher, letter to Energy Secretary Hazel O'Leary, July 10, 1996; copy in author's possession.

page 82: "fringe critics who ... prefer . . . : Rep. George E. Brown, Jr., "Mythmakers and Soothsayers: The Science and Politics of Global Change," speech delivered to the NATO/Duke University School of the Environment Workshop on Global Change Integrated Risk Assessment, Duke University, October 15, 1995.

page 82: a succession of uncontrolled fires . . . : "Mongolian Airborne Called to Fight Blazes," *Boston Globe,* May 5, 1996; "Mongolia Is Devastated by Fires of Epic Scope," *New York Times,* June 9, 1996.

CHAPTER FOUR: THE CHANGING CLIMATE OF BUSINESS

page 85: "wants you to believe that . . . : Michael Marvin, interview by author, May 30, 1996.

page 86: "Some ... organizations have, on occasion . . . : J. S. Jennings, "Future Sustainable Energy Supply," address to the Sixteenth World Energy Council Congress, Tokyo, Japan, October 9, 1995.

page 86: And in October 1996 BP America . . . : Ehsan Masood, "Companies Cool to Tactics of Global Warming Lobby," *Nature,* vol. 383 (October 10, 1996).

page 87: a report recommending that banks . . . : Mark Mansley, *Long Term Financial Risks to the Carbon Fuel Industry from Climate Change* (London: Delphi Group, 1995).

page 87: H. R. Kaufmann declared . . . : Quotes from Kaufmann and Nutter from: "Storm Warnings: Climate Change Hits the Insurance Industry," *World Watch,* November/December 1994.

page 88: In the 1980s insurance payouts . . . : Insurance industry data from Munich Re/Münchener Rückversicherungs-Gesellschaft, in Christopher Flavin and Odil Tunali, *Climate of Hope: New Strategies for Stabilizing the World's Atmosphere* (Worldwatch Institute, June 1996).

page 88: In 1995 fourteen of the world's . . . : "Insurers Lobby for Environment," Reuters, March 29, 1995.

page 88: In October 1996, one of the largest . . . : Kaj Ahlmann letter to Vice President Al Gore, October 2, 1996; copy in author's possession.

page 89: "Insurers understand that environmental risks . . . : United Nations Environment Programme, "Leading Insurance Firms Take Environmental Pledge," press release, November 23, 1995.

page 89: "It is not easy to get people . . . : Franklin Nutter, interview by author, March 20, 1996.

page 89: the number of federally certified disasters . . . : John McShane, interview by author, April 11, 1996.

pages 91–93: "The fundamental science of global warming . . . : Kevin Fay, interview by author, March 28, 1996.

pages 93–94: Quotations from Sven Hansen, Hilary Thompson, and Kaspar Mueller come from Jeremy Leggett and Greenpeace U.K., eds., *Climate Change and the Financial Sector: The Emerging Threat—The Solar Solution* (Munich:Gerling Akademia Verlag, 1996). This book contains the proceedings of the Climate Summit of representatives of financial industries, Berlin, 1995.

pages 94–95: "civil war in the energy industry" . . . : Jeremy Leggett, interview by author, March 9, 1996.

pages 95–96: "The myth is that environmentally . . . : Mark Marvin, interview by author, May 30, 1996.

page 97: "These changes will happen anyway . . . : Kirk Brown, interview by author, March 29, 1996.

page 98: "That acceleration of renewable energy . . . : Alden Meyer, interview by author, March 19, 1996.

page 98: that number is actually on the order of . . . : Doug Koplow, "Energy Subsidies and the Environment," in *Subsidies and Environment: Exploring the Linkages* (Organization for Economic Cooperation and Development, 1996).

pages 99–100: "already imposed caps on how much insurance . . . : Katherine Raupp, interview by author, May 6, 1996.

pages 100–101: the only plan for an energy transition that . . . : John Schlaes, interview by author, May 30, 1996.

page 102: "We do not believe the United States . . . : "Six Democratic Senators Urge Caution in Climate Talks," Reuters, July 18, 1996.

page 104: drought and hot, dry winds . . . : "Fierce Struggle in Southwest Tinderbox," *New York Times,* May 7, 1996.

page 104: a state of disaster was declared . . . : "Australian Floods Leave One Dead, Three Missing," Reuters, May 6, 1996.

CHAPTER FIVE: AFTER RIO: THE SWAMP OF DIPLOMACY

page 107: In June 1992, 132 heads of state . . . : Portions of this chapter are based on author's interviews with diplomats, delegates, and officials from several nations who spoke only on the condition that they not be identified.

page 111: "This is an essential first step . . . : "Gummer Sounds Alarm Bells on Global Warming," *The Times* (London), July 18, 1996.

page 112: The Australian officials refused . . . : "US Greenhouse Switch Leaves Us in the Cold," *Sydney Herald* [Australia], May 15, 1996.

page 112: "Two hundred years after the Industrial . . . : "China's Inevitable Dilemma: Coal Equals Growth," *New York Times,* November 29, 1995.

pages 113–114: "The United States is one of . . . : Anil Agarwal, interview by author, April 5, 1996.

page 115: Asian greenhouse emissions . . . : "CO_2 Emissions Will Jump in Asia By 2025," *Daily Yomiuri* (Tokyo), May 25, 1996.

page 115: "Sea levels will rise up to three feet . . . : "Scientists Predict Coastal Havoc," UPI, May 28, 1996.

page 115: In an official 1996 white paper . . . : "China Calls for Environmental Help," UPI, June 6, 1996.

page 116: several developing countries are already . . . : Some of the examples are taken from Patrick Keegan, Bob Price, Caroline Hazard, Cathryn Stillman, and Greg Mock, "Developing Country Greenhouse Gas Mitigation: Experience with Energy Efficiency and Renewable Energy Policies and Programs," a paper presented to the Climate Change Analysis Workshop, June 1996.

page 116: a dozen solar panels and streetlights . . . : "New Power Sources Tried in Rural Cuba," *Los Angeles Times* service, June 23, 1996.

pages 117–118: "Certainly we have strong economic priorities . . . : Antonio LaVina, interview by author, May 4, 1996.

page 118: "environmental colonialism": Anil Agarwal and Sunita Narain, *Global Warming in an Unequal World* (New Delhi:Centre for Science and Environment, 1992).

page 119: Pearlman is registered as a representative . . . : Exhibits and amendments to registration statement pursuant to the Foreign Agents Registration Act, submitted by Patton, Boggs and Blow to the U.S. Department of Justice; copy in author's possession.

page 119: According to its IRS charter . . . : IRS forms 990 and 1023 of the Climate Council, obtained through a Freedom of Information request to the Internal Revenue Service.

page 120: "The IPCC has been heavily politicized . . . : Donald Pearlman, telephone interview by author, August 7, 1995.

page 121: "Business representatives present at this meeting . . . : *ECO, The CAN Climate Negotiations Newsletter,* vol. 1, March 4, 1996.

page 122: "there were deliberate attempts to obfuscate . . . : Kevin Trenberth, *Report to the IPPC Working Group I,* December 8, 1995; letter from Trenberth to Sir John Houghton, February 7, 1996; copies of both in author's possession.

page 123: In November 1995, after the Saudis . . . : "Climate panel confirms human role in warming, fights off oil states," *Nature,* vol. 378 (Dec. 7, 1995).

page 123: "This is the work of two thousand scientists . . . : "Global Warming Divide Deepens," *Guardian,* July 17, 1996.

pages 124–126: Quotations from Wirth, Schlaes, and Linderman are drawn from various newspaper and wire service accounts of the Geneva conference.

page 125: "At an average cost of . . . : *Eco 2,* The Climate Action Network Climate Negotiations Newsletter, March 5, 1996.

page 128: "the appalling state of compliance . . . : Kilaparti Ramakrishna and Andrew M. Deutz, "Ecological Considerations in the Setting of Quantified Emissions Limitation and Reduction Objectives: Implications of the Ultimate Objective of the U.N. Framework Convention on Climate Change," paper prepared for the Climate Change Analysis Workshop, June 6, 1996.

page 130: "likely to cause severe economic dislocations. . . . : Mobil advertisement in *New York Times,* July 25, 1996.

page 130: As the economic journalist Robert Kuttner . . . : "Global Competitiveness and Human Development: Allies or Adversaries?", the 1996 Paul G. Hoffman Lecture by Robert L. Kottner, delivered to the United Nations Development Programme, November 1, 1996, New York.

page 131: "We need to explain why . . . : William Ruckelshaus, interview by author, May 9, 1996.

page 131: "Ultimately, the question goes beyond . . . : Antonia LaVina, interview by author, May 4, 1996.

page 132: "more than 100 people were killed in heavy flooding . . . : "Floods Kill 100 Afghans; Officials Fear More to Come," *New York Times,* April 23, 1996.

page 132: 17 inches of rain fell in twenty-four hours . . . : "Rain of Biblical Proportions Pours Out of Midwest Skies," *New York Times,* July 20, 1996.

page 132: a torrent of mud and rock . . . : "Flash Floods in Spanish Pyrenees Kills Scores," *New York Times,* August 9, 1996.

CHAPTER SIX: HEADLINES FROM THE PLANET

page 136: Ocean Warming Creates Pacific Wasteland: "Climatic Warming and the Decline of Zooplankton in the California Current," Dean Roemmich and John McGowan, *Science,* vol. 267 (March 3, 1995). McGowan quoted in "Warming of Seas Creates a Pacific Wasteland Off San Diego," *New York Times,* March 5, 1995.

page 136: over the past few decades a significant warming . . . : "Climatic Warming and the Decline of Zooplankton in the California Current," Dean Roemmich and John McGowan, *Science,* vol. 267 (March 3, 1995).

page 137: Small Temperature Rise Fuels Migrations of Sea Animals: "Study Suggests Some Sea Creatures Responding to Changing Climate," Associated Press, February 3, 1995.

page 137: Baxter, co-author of a study . . . : "Climate Related, Long-Term Faunal Changes in a California Rocky Intertidal Community," J. P. Barry, et al., *Science,* February 3, 1995.

page 138: "Among the repercussions . . . : *Global Warming and Biological Diversity,* Robert L. Peters and Thomas E. Lovejoy, eds., (Yale University Press, 1992).

page 138: Butterfly Study Confirms Warming-Driven Migrations: "Western Butterfly Shifting North as Global Climate Warms," *New York Times,* September 3, 1996; Camille Parmesan, "Climate and Species Range," *Nature,* vol. 382 (August 29, 1996).

page 139: Melting of the World's Glaciers Accelerates: Mark Meier, *Earth,* 1995.

page 139: One indication is that the volume of the world's half-million small glaciers . . . : Accounts of two studies by glaciologist Mark Meier, et al., of the University of Colorado, presented at the International Union of Geodesy and Geophysics Conference, 1995, in "Earth's Shrinking Glaciers," *Earth,* 1995.

page 139: Comparing current measurements to photos . . . : James Peterson and Geoffrey Hope (researchers quoted in article), "Meltdown Warning as Tropical Glaciers Trickle Away," *New Scientist,* June 24, 1995.

page 139: Strong indications of warming have been found . . . : Lonnie G. Thompson and Ellen Mosley-Thompson, "Late Glacial Stage and Holocene Tropical Ice Core Records from Huascaran, Peru," *Science,* vol. 269 (July 7, 1995).

page 140: ice cores "from diverse locations . . . : L. G. Thompson, and E. Mosley-Thompson, "Glaciological Evidence for Recent Warming at High Elevations," 76th American Meteorological Society Annual Meeting, *Symposium on Environmental Applications.* Preprint in author's possession; L. G. Thompson, E. Mosley-Thompson, M. Davis, P. N. Lin, T. Yao, M. Dyurgerov, and J. Dai, "Recent Warming: Ice Core Evidence from Tropical Ice Cores with Emphasis on Central Asia," *Global and Planetary Change,* vol. 7 (1993).

page 140: The retreat of tropical glaciers correlates closely . . . : Henry F. Diaz and Nicholas E. Graham, "Recent Changes in Tropical Freezing Heights and the Role of Sea Surface Temperature," *Nature,* vol. 383 (September 12, 1996).

page 140: after an initial spurt, the trees' growth rate flattened . . . : Gary Taubes, "Is a Warmer Climate Wilting the Forests of the North?" *Science,* vol. 267 (March 15, 1995).

page 141: The warming appears to be stressing . . . : Gary Taubes, "Is a Warmer Climate Wilting the Forests of the North?" *Science,* vol. 267 (March 15, 1995).

page 141: North of the Canadian forests, a series of boreholes in Alaska revealed soil temperature increases of . . . : David Deming, "Climatic Warming in North America: Analysis of Borehole Temperatures," *Science,* vol. 268 (June 16, 1995).

page 142: Deming's findings parallel a discovery the previous December that a deep layer of the Arctic Ocean has warmed . . . : Antonio Regalado, "Listen Up! The World's Oceans May Be Starting to Warm," *Science,* vol. 268 (June 9, 1995).

page 142: Heat-Enhancing Vapor Increases . . . : S. J. Oltmans, and D. J. Hofmann, "Increase in Lower-Stratospheric Water Vapour at a Mid-latitude Northern Hemisphere Site from 1981 to 1994," *Nature,* vol. 374 (March 9, 1995).

page 143: Recent El Niño a One-in-Two-Thousand-Year Event: Kevin Trenberth and Timothy Hoar, "The 1990–1995 El Niño–Southern Oscillation Event: Longest on Record," *Geophysical Research Letters,* vol. 23, no. 1 (January 1, 1996).

page 144: "We have had a whole series . . . : Barry Parker, quoted in *New Scientist,* January 16, 1993.

page 144: "there is evidence we seem to be . . . : John Gould, quoted in *New Scientist,* January 16, 1993.

page 144: U.S. Wheatfields Could Be Deserts in a Decade: William K. Stevens, "Great Plains or Great Desert? A Sea of Dunes Lies in Wait," *New York Times,* May 28, 1996.

page 145: Desert Conditions Spreading in Southern Europe: John Thornes, "Deserts on Our Doorstep," *New Scientist,* July 6, 1996.

pages 145–146: "our measurements of the increase in seasonal changes . . . : Charles Keeling, "Increased Activity of Northern Vegetation Inferred from Atmospheric CO_2 Measurements," *Nature,* vol. 382 (July 11, 1996).

page 146: "The drawdown in carbon dioxide . . . : Charles Keeling, quoted in *Science News,* July 13, 1996.

pages 146–147: Scientists Discover Further Disintegration . . . : Vaughan, Zwally, and del Valle, quoted in "Melting Ice Stirs Fear on Warming," *Boston Globe,* January 25, 1996; "Ice Shelves Melting as Forecast, but Disaster Script Is in Doubt," *New York Times,* January 30, 1996.

pages 146–147: surface temperatures in the Antarctic Peninsula . . . : D. G. Vaughan and C. S. M. Doake, "Recent Atmospheric Warming And Retreat of Ice Shelves on the Antarctic Peninsula," *Nature,* January 25, 1996.

pages 147–150: Portions of the section on infectious diseases are taken from author's interview with Dr. Paul R. Epstein of the Harvard School of Public Health's Working Group on Emerging Diseases. Epstein is a principal author of the chapter on health effects and climate change for the WHO/WMO/UNEP intergovernmental assessment and a member of the U.S. government's Working Group on Emerging and Re-Emerging Infectious Diseases.

page 147: In December 1995 the IPCC working group on climate impacts . . . : *Scientific Technical Analysis Of Impacts, Adaptations, and Mitigation of Climate Change,* IPCC Working Group II, December 1995.

page 147: there is a correlation between climate change . . . : "Global Climate Change and Emerging Infectious Diseases," *Journal of the American Medical Association,* January 16, 1996.

page 148: a series of similar reports in 1994 in *The Lancet* . . . : David Sharp and Paul R. Epstein, eds., *Health and Climate Change,* The Lancet Ltd.: (Devonshire Press, London, 1994).

page 148: "we do not have the usual option . . . : A. J. McMichael, A. Haines, R. Slooff, and S. Kovats, eds., *Climate Change and Health: An Assessment Prepared by a Task Group on Behalf of the World Health Organization, the World Meteorological Organization and the United Nations Environmental Programme,* World Health Organization, Geneva, Switzerland, 1996.

page 148: As we have seen, the *Aedes aegypti* mosquito . . . : David Sharp and Paul R. Epstein, eds., *Health and Climate Change,* The Lancet Ltd., (London: Devonshire Press, 1994).

page 148: Recently, however, this mosquito has been reported . . . : Paul Epstein, interview with author.

page 148: "epidemic potential of the mosquito population": Willem J. M. Martens et al., "Potential Impact of Global Climate Change on Malaria Risk," *Environmental Health Perspectives,* May 1995.

page 149: attributes the spread of the disease to two factors: A. Huq, R. R. Cowell, et al., "Coexistence of *Vibrio cholerae* 01 and 0139 Bengal in plankton in Bangladesh," *Lancet,* vol. 345, 1995.

page 151: more than 330 people died . . . : "Malaria Kills 30 in Yemen, Hits Governorate," Reuters, August 3, 1996.

CHAPTER SEVEN: THE COMING PERMANENT STATE OF EMERGENCY

pages 153–155: "Long before the systems of the planet buckle, . . . : William Ruckelshaus, interview by author, May 9, 1996; Henry Kendall, interview by author, April 1, 1996.

page 155: "the Green Revolution has been . . . : Henry W. Kendall and David Pimentel, "Constraints on the Expansion of the Global Food Supply," *Ambio,* vol. 23 (May 3, 1994). *Ambio* is the journal of the Royal Swedish Academy of Sciences.

page 159: the lower reaches of the Yellow River . . . : "China's Fickle Rivers: Dry Farms, Needy Industry Bring a Water Crisis," *New York Times,* May 23, 1996.

pages 159–60: "Vulnerability to climate change . . . : Cynthia Rosenzweig and Daniel Hillel, "Potential Impacts of Climate Change on Agriculture and Food Supply," *Consequences,* vol. 1, no. 2 (Summer 1995). *Consequences* is a publication of NASA, NOAA, and the National Science Foundation.

pages 161–62: areas of the world that are both the poorest . . . : Norman Myers and Jennifer Kent, *Environmental Exodus: An Emergent Crisis in the Global Area* (Washington, D.C.: Climate Institute, 1995).

pages 163–64: "Unexpected 'surprises' may well accompany . . . : Rosenzweig and Hillel, "Potential Impacts of Climate Change on Agriculture and Food Supply," *Consequences,* vol. 1, no. 2 (Summer 1995).

page 163: the wheat fields could turn—in a decade—into a vast desert . . . : William K. Stevens, "Great Plains or Great Desert? A Sea of Dunes Lies in Wait," *New York Times,* May 28, 1996.

page 165: "The World Trade Center was easy . . . : Norman Myers, interview by author, April 5, 1996.

page 166: FEMA officials say they cannot consider . . . : John McShane, interview by author, April 11, 1996.

page 167: "This is democracy's time on the world stage . . . : William Ruckelshaus, interview by author, May 9, 1996.

page 169: parts of Kansas and Oklahoma were suffering . . . : "Worst Drought Since '30s Grips Plains," *New York Times,* May 20, 1996.

page 169: On November 8, 1996, the worst cyclone of the century . . . : "Storm Toll Tops 1,000 in India," *The Boston Globe,* November 9, 1996; Reuters, November 11, 1996.

CHAPTER EIGHT: ONE PATHWAY TO A FUTURE

page 172: The experience of Dr. Daniel Goodenough . . . : Remarks by Daniel Goodenough at Harvard Medical School, June 1996.

page 174: what writer . . . : Tom Athanasiou, *Divided Planet: The Ecology of Rich and Poor* (Boston:Little, Brown, and Company, 1996).

page 175: "It is no exaggeration . . . : Jonathan Weiner, *The Beak of the Finch* (New York:Alfred A. Knopf, 1994).

page 177: some of the highlights of 1995, as recounted by the Worldwatch Institute . . . : "Vital Signs," Worldwatch Institute, May 19, 1996, quoted in " '95 Best, Worst of Times, Global Study Says," *Boston Globe,* May 19, 1996.

page 178: As energy analysts Joseph Romm . . . : Joseph J. Romm and Charles B. Curtis, "Mideast Oil Forever?" *Atlantic Monthly,* April 1996.

page 179: Christopher Flavin of the Worldwatch Institute . . . : Christopher Flavin, "Power Shock: The Next Energy Revolution," *World Watch,* January–February 1996.

page 179: the design for a new nonpolluting "hypercar" . . . : Amory Lovins, presentation at Tufts University, May 7, 1996.

page 179: the economics panel of the IPCC has identified . . . : IPCC Working Group III, *Summary for Policy Makers: Second Assessment Report,* November 1995.

page 180: the current "pattern of subsidies . . . : Doug Koplow, "Energy Subsidies and the Environment," in *Subsidies and Environment: Exploring the Linkages* (Organization for Economic Cooperation and Development, 1996).

page 181: "It is not as hard as it might seem . . . : Michael Marvin, interview by author, May 30, 1996.

page 181: "There is no point of contact . . . : Herman Daly, quoted in "Fitfully, Remedies for Planet Emerge," *Boston Globe,* June 2, 1992.

page 182: "the Gross Domestic Product" portrays disaster . . . : Clifford Cobb, Ted Halstead, and Jonathan Rowe, "If the Economy Is Up, Why Is America Down?" *Atlantic Monthly,* October 1995.

page 182: The three economists . . . as well as economist Daly . . . : Herman E.

Daly and John B. Cobb, Jr., *For The Common Good* (Boston:Beacon Press, 1994).

page 184: They don't understand what economics journalist Robert Kutter calls . . . : Robert Kutter, *Everything For Sale: The Virtues and Limits of Markets* (Alfred A. Knopf and Twentieth Century Fund, 1997).

page 189: for every million dollars spent on oil and gas exploration . . . : Scott Sklar, "Renewable Energy Technologies," paper presented at the Climate Change Analysis Workshop, June 6–7 1996.

page 194: On even the calmest days, a limb . . . : "Termites Haunt, and Topple, Mighty Oaks in Leafy New Orleans," *New York Times,* June 30, 1996.

APPENDIX REFERENCES

The following works are cited in articles reproduced in the Appendix:

Angell, J.K., et al., "The Effect Of Moisture On Layer Thicknesses Used to Monitor Global Temperatures." *Journal Of Climate,* 1994 Feb., V7, N2:304–308, (and updates) from *Trends 93,* U.S. Department of Energy, 636–672.

Balling, R. *The Heated Debate.* San Francisco: Pacific Research Institute for Public Policy, 1992.

Barnett, T., and M. Schlesinger. 1987. *Journal of Geophysical Research* 92:14772–80.

Hansen, J. H., H. Wilson, M. Sato, R. Ruedy, K. Shah, and E. Hansen. 1995. "Satellite and Surface Temperature Data at Odds?" *Climate Change* 30(1):103–17.

Hasselmann, K., L. Bengtsson, U. Cubasch, G. C. Hegerl, H. Rodhe, E. Roeckner, H. V. Storch and R. Voss. 1995. "Detection of Anthropogenic Climate Change Using A Fingerprint Method." Max Planck Institut Fur Meteorologie, Report No. 168.

Hasselmann, K. 1979. "On The Signal-To-Noise Problem in Atmospheric Response Studies." In: *Meteorology of Tropical Oceans,* D. B. Shaw ed., 251–259. London: Royal Meteorological Society.

Hegerl et al. 1996. "Detecting Anthropogenic Climate Change with an Optical Fingerprint Method." *Journal of Climate.*

Hegerl et al. (1995), "A Climate Change Simulation Starting From 1935." *Climate Dynamics,* 1995 March, V11 N2:71–84.

Climate Change: The IPCC Scientific Assessment. J. T. Houghton, G. J. Jenkins and J. J. Ephraums, eds., 1990. Cambridge, U.K.: Cambridge University Press.

Climate Change 1992: The Supplementary Reports to the IPCC Scientific Assess-

ment. J. T. Houghton, B. A. Callender, and S. K. Varney, eds. 1992. Cambridge, U.K.: Cambridge University Press.

Climate Change 1995: The Science of Climate Change, Contribution of Working Group I to the Second Assessment Report of the Intergovernmental Panel on Climate Change. J. T. Houghton, L. G. M. Filho, B. A. Callander, N. Harris, A. Kattenberg, and K. Maskell, eds. 1996. New York: Cambridge University Press.

Intergovernmental Panel on Climate Change (IPCC), Working Group I Second Assessment Report.

Karl, T. R., R. W. Knight, and N. Plummer. 1995. "Trends in High-Frequency Climate Variability in the 20th Century." *Nature* 337:217–20.

Karoly et al. 1994. "An Example of Fingerprint Detection of Greenhouse Climate Change." *Climate Dynamics* 10:97–105.

Kattenberg, A., F. Giorgi, H. Grassi, G. A. Meehl, J. F. B. Mitchell, R. J. Stouffer, T. Tokioka, A. J. Weaver, and T. M. L. Wigley. 1996. "Climate Models: Projections of Future Climate." In Houghton et al., *Climate Change 1995.*

MacCracken, M. 1993. Review of Balling's *Heated Debate. Bulletin of the American Meteorological Society* 74 (9):1752–54.

Manabe, S., R. J. Stouffer, M. J. Spelman, and K. Bryan. 1991. "Transient Responses of a Coupled Ocean-atmosphere Model to Gradual Changes of Atmospheric CO_2. Part I: Annual Mean Response" *Journal of Climate* 4:785–818.

Michaels, P. J., and D. E. Stooksbury. 1992. "Global Warming: A Reduced Threat?" *Bulletin of the American Meteorological Society* 73:1563–77.

Michaels, P. J. et al; "Predicted and Observed Long Night and Day Temperature Trends." *Atmospheric Research,* 1995 July, V37 N1-3:257–266.

Mitchell, J. F. B., T. C. Johns, J. M. Gregory, and S. F. B. Tett. 1995. "Climate Response to Increasing Levels of Greenhouse Gases and Sulphate Aerosols." *Nature* 376:501–04.

Oort, A. H. and H. Lui. "Upper-Air Temperature Trends Over the Globe, 1958–1989." *Journal of Climate* 6:292–307.

Ramaswamy, V., M. D. Schwarzkopf, and W. J. Rendel. 1996. "An Unanticipated Climate Change in the Global Lower Stratosphere Due to Ozone Depletion." (Submitted to *Nature.*)

Roeckner, E., T. Siebert, and J. Feichter. 1995. "Climate Response to Anthropogenic Sulphate Forcing Simulated with a General Circulation Model." In *Aerosol Forcing of Climate,* R. Charlson and J. Heintzenberg, eds., 349–62. New York: John Wiley and Sons.

Santer, B. D. 1995[a]. "Towards the Detection and Attribution of An Anthropogenic Effect on Climate." *Climate Dynamics* 12:79–100.

Santer, B. D., et al. 1996. "A Search for Human Influences on the Thermal Structure of the Atmosphere." *Nature* 382.

Santer, B. D., T. M. L. Wigley, and P. D. Jones. 1993. "Correlation Methods in Fingerprint Detection Studies." *Climate Dynamics* 8:265–76.

Santer, B. D., T. M. L. Wigley, T. P. Barnett, and E. Anyamba. 1996. Detection of Climate Change and Attribution of Causes." In Houghton et al., *Climate Change 1995.*

Singer, S. Fred. 1996. Letter to the editor. *Science.*

———. 1996. Letter to the editor. *Wall Street Journal,* July 25, 1996.

Taylor, K. E. and J. E. Penner. 1994. "Anthropogenic Aerosols and Climate Change." *Nature* 369:734–36.

Wigley, T. M. L. 1989. "Possible Climatic Change Due to SO_2-derived Cloud Condensation Nuclei." *Nature* 339:365–67.

Wigley, T. M. L. and T. P. Barnett. 1990. "Detection of the Greenhouse Effect in the Observations." In Houghton, *IPCC Scientific Assessment,* pp. 239–55.

Wigley, T. M. L. and S. C. B. Raper. 1987. "Thermal Expansion of Sea Water Associated with Global Warming." *Nature* 330:127–31.

———. 1991. "Detection of the Enhanced Greenhouse Effect on Climate." In *Climate Change: Science, Impacts, and Policy,* J. Jaeger and H. L. Ferguson, eds., 231–42. Cambridge, U.K.:Cambridge University Press.

———. 1992. Implications for climate and sea level of revised IPCC emissions scenarios. *Nature* 357: 293–300.

Wigley, T. M. L., B. Santer, J. F. B. Mitchell, and R. J. Charlson. 1996. Letter to the editor. *Science.*

INDEX